The Age of Mammals

The Age of Mammals

Björn Kurtén

Columbia University Press
New York

Published in Great Britain in 1971
by Weidenfeld and Nicolson

Library of Congress Catalog Card Number: 79–177479

ISBN 0-231-03647-7

Printed in the United States of America
10 9 8 7 6 5 4 3 2

For Solveig

Contents

Figures

Plates

Preface

AT THE beginning of the nineteenth century, the French scientist Cuvier astonished the world by his discoveries of the remains of huge, extinct creatures in the Paris area. Those fantastic beings, whose fossil bones were found in profusion in the Gypse de Montmartre and at other sites, had clearly been mammals, but mammals of kinds never seen by man. Cuvier's work, and that of his successors – men like Gaudry in France, Owen in England, Leidy, Marsh and Cope in North America, and the Ameghino brothers in South America – showed that the age we live in was preceded by an Age of Mammals: an era in earth history before the coming of man.

Thousands of scientists have since then continued to collect and correlate facts and details – geologists, paleontologists, oceanographers, geophysicists, and many others – and gradually a picture of this complex and remarkable era is emerging. The present book is an attempt to review it. It does not deal only with the mammals, although they certainly are the main actors on the scene, but also with other forms of life, and the stage on which they appeared. The changing geography and the changing climates are as much a part of the story as the changing fauna and flora. And the time-scale of the story, which is now known to us, would have staggered Cuvier, who wrote at a time when the age of the earth was believed to be a few thousand years. We now know that the Age of Mammals lasted for more than sixty million years, and yet it is but a fraction of geological time as a whole.

In 1910, H. F. Osborn published his renowned book *The Age of Mammals in Europe, Asia and North America*. It was a rather technical account, summarising information then available. In the present book my aim has been to present the story of the Age of Mammals in a concise and non-technical manner, and at the same time to try and help the reader to sense the magnitude of its scale; and perhaps also to ponder the responsibility of man for this earth of ours, that has supported such an abundance of life during ages longer than anything we can comprehend. It seems to me that such an historical perspective is one of the most vital messages that paleontology has for modern man, steeping his surroundings in poison and filth, threatening to break the chain of life forged during several billion years.

The book was planned as an independent sequel to E. H. Colbert's *The Age of Reptiles*; hence its name. It was embarked upon in 1963, at the suggestion of Richard Carrington, but because of various other commitments seven years were to elapse before it was finished. I am grateful to Richard Carrington and Julian Shuckburgh for constant help and encouragement, and most particularly to Sonia Cole who edited the script and made valuable suggestions and corrections. My colleagues Peter Robinson, Craig C. Black and Miguel Crusafont Pairó kindly read an early draft of the script and made numerous helpful comments.

It was decided to illustrate the book solely with reconstructed scenes and maps depicting the living past, and I have been most fortunate in securing the talent and competence of Margaret Lambert for this task. Working independently, she has approached the theme from the artist's point of view, and so produced a pictorial version of the Age of Mammals as a complement to the text.

My work has been supported by grants from the University of Helsinki, the Finnish State Commission for Natural Science, and the Societas Scientiarum Fennica. I wish to record my gratitude to all the persons and institutions mentioned.

1970 Björn Kurtén

Chapter 1

Fossils and the origin of mammals

> *All* What, never?
> *Captain* No, never!
> *All* What, *never*?
> *Captain* Hardly ever!
> W. S. Gilbert, *H.M.S. Pinafore*.

'NEVER', TO a statistician, means that the probability of a given event is zero. 'Hardly ever', on the other hand, refers to an event with a probability very low but still greater than zero – say 0·000001, meaning one chance in a million. Surely no rational gambler would take a chance on that. But we are perhaps not as rational as we like to think, and many of us in fact buy lottery and sweepstake tickets against greater odds than that.

If the sweepstake gambler seems irrational to us, then who would dream of a science based on the millionth chance? Yet this is, for all practical purposes, the basis of vertebrate paleontology – the study of the back-boned animals of the past. In most cases, our sole source of knowledge is the fossil bones that have been preserved in the earth. But when an animal dies in nature, its remains are usually destroyed within a short time. The meat rots away or is devoured by scavenging birds and mammals; the bones are broken up by specialists like the hyena and the glutton; and the remaining fragments disintegrate under the erosive effects of sunshine, frost, rainfall and drought. Dust returns to dust. Even the mighty bones of an elephant, left on the veldt, will eventually vanish, leaving no trace of that great carcass.

Perhaps once in a million times, however, the skeleton or part of it becomes buried in newly-formed sediment, so escaping destruction and preserving for posterity the record of this animal. It could be an antelope that died in a dry stream-bed or wadi; the next flood then swept its clean-picked bones along and deposited them in the gravels. Or it could be a cave bear dying peacefully in its winter sleep, in the darkness of its cavern. In the spring there would be an uproarious funeral staged by foxes and gluttons and the big bones and teeth would be left, perhaps to become imbedded in the gradually accumulating cave sediments. The odds on fossilisation are better in some types of environment, worse in others; yet they are nearly always very long and this may seem a slender basis for a science.

Yet there is an all-important difference between 'never' and 'hardly ever', between the impossible and the very improbable. Whenever the probability is finite, even just a chance in a million, the event will become probable if the number of trials is great enough. If the species numbers one million individuals in each generation, the odds that at least one of them will be fossilised are almost two to one; and in ten generations the chances will rise to 0·99995, twenty thousand to one in favour of fossilisation. Any gambler would regard this as a sure thing.

Still, although fossilisation may be a certainty for many species of vertebrate animals, there remain many hazards before the fossil appears where it can give us information – in the laboratory of the scientist. The deposit that held the bones may have been attacked by erosion and destroyed with all its contained fossils. It may now be lying deeply buried under later sediments, out of the reach of excavators, except in the case of a lucky shot in the course of mining or deep boring. It may have been so altered by pressure and heat under the earth that all traces of its organic contents are gone. Or it may be in some region that the field paleontologist has not yet reached.

After finding the bones, the paleontologist faces the task of taking them out of the sediment with as little damage to them as possible and bringing them back to the laboratory, together with complete records of the circumstances of the finds. This is only the beginning of the job. Next comes the task of identifying the animal to which the bones belonged. It may turn out to be a species already known to science, which indeed is generally the case. But sometimes the paleontologist is thrilled by the discovery of a new species hitherto unknown to science. It will then be necessary to establish its relationship to known species and to give it a name.

Let us assume that the fossil is a mammalian tooth like that shown in Figure 1. Its high, prism-like shape and the enamel pattern on its grinding surface show that it is an upper cheek tooth of a highly evolved member of the horse family. Details like the intense complication of the enamel folds and the isolated loop called the protocone identify it as a species of the extinct genus *Hipparion*. Identification of the exact species may be difficult on the basis of a single tooth, but in this case we also have other skeletal material and the sum of characters enables us to identify it with the well-known species *Hipparion sitifense*.

Once the fossil has been identified, we may try to form some idea of how the animal looked in the flesh and what its mode of life was like. The

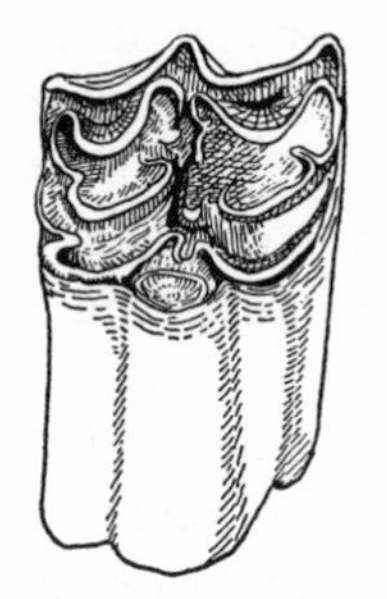

Figure 1. Upper molar of the small three-toed horse *Hipparion sitifense* from the vicinity of Teruel, Spain, shows complicated enamel pattern typical of advanced horses; isolated inner loop (in front), or protocone, is characteristic of *Hipparion*.

feeding habits, for instance, are almost always reflected in the dentition. Fortunately, teeth are particularly common as fossils, because their hard enamel keeps them from falling apart even in conditions where other bones may disintegrate. Indeed, some extinct species are known from teeth only.

Grass-eaters like the bison and the horse (also *Hipparion*) have high-crowned cheek teeth with a complicated enamel pattern, which develops an efficient grinding surface as it wears down. In leaf-eating or browsing woodland herbivores like deer and giraffes, the tooth crowns are lower and the cusps remain separated to form a chopping mechanism. Flesh-eaters or carnivores like the lion and the stoat have cheek teeth forming elongated, shearing blades; if they specialise in grinding bone like the hyena, the teeth become heavy conical structures adapted for bone-crushing. Insect-eaters or insectivores like the shrews and hedgehogs have cheek teeth with numerous sharply-pointed cusps. Omnivores like man and the bear also have many-cusped teeth, but the cusps are low and blunt.

An entirely different avenue for the study of food preferences of extinct animals has recently been explored. It has been shown that the strontium content of bone is related to the food consumed. The concentration of strontium tends to be especially high in succulent herbaceous vegetation, while it is lower in grasses and still lower in meat. In a sample of Pliocene vertebrate fossils from North America the strontium content was found to agree with expectation, as based on the dentition of the animals and other inferences about their mode of life. One of these forms was a type of horse (*Hypohippus*) which has always been regarded as a browsing forest dweller, because of its low-crowned cheek teeth and other characters. The strontium concentration in its bones was found to be 630 parts per million, the same as in the tortoise *Testudo* which we know to be a browser. On the other hand, concentrations found in the remains of probably grass-eating forms like the steppe horse *Pliohippus* and the pronghorn-like *Merycodus* were much lower (523–552 ppm). Carnivore bones gave still lower values (only 477 ppm). Here is a new and unexpected tool for the study of the past. We may be sure that others will be invented in future.

The bones of the feet are usually very massive and durable in structure and so are frequently found. They give important information on the posture and gait of the animal. On rare occasions the fossil hunter may be rewarded by the find of a whole skeleton or the greater part of one. Then a careful study of the various joints and the muscle attachments may enable us to make a restoration of the entire animal as it probably looked in life. In the case of *Hipparion*, there are plenty of finds on which a life restoration may be based. It was a relatively small horse, generally resembling a present-day donkey or zebra, but with three toes on each foot instead of one as in all living members of the horse family. Restorations often show *Hipparion* with zebra-like stripes but this, like the shape of the ears, the mane, and the hair whisk at the end of the tail, is of course hypothetical.

The time has come to find out about the age of our fossil. This is done by comparing it, and the other fossils found in the same stratum (as well

Figure 2. Life restoration of the pony-sized, three-toed horse *Hipparion*, a common animal in Eurasia throughout the Pliocene epoch. Complete skeletons of this animal have been found at Höwenegg in southern Germany.

as in underlying and overlying strata), with the geological succession of fossils that has been compiled by the combination of a great number of local sequences. Our *Hipparion sitifense*, for instance, probably lived during the middle part of the Pliocene epoch (one of the epochs of the Tertiary period; see below), because the known fossils of the species in Europe and North Africa all seem to date from that time. Study of associated fossils from the same beds will help to confirm the dating and make it more precise.

This is relative dating: the age of the fossil is determined in relation to other events, not in terms of so many years. In fact the geological history of the earth was established as a sequence of events long before they could be dated in years.

That part of geological time during which abundant fossil remains have been deposited is divided into three great eras: the Paleozoic, Mesozoic, and Cenozoic. Let us take a brief look at the history of life during these eras.

The eras are divided into periods, of which there are six (or, some say, seven) in the Paleozoic. From the earliest period, the Cambrian, we know no vertebrate animals. They make their appearance in the second period, the Ordovician; unfortunately the material at hand is very poor, but it does indicate the presence of jawless 'fishes' related to the living lampreys. In the third period, the Silurian, jawless vertebrates are found in some abundance, while the fourth or Devonian period saw such a profusion of primitive fishes that it has been called the Age of Fishes. Among them were the lobe-finned fishes or crossopterygians that were to give rise to land-living vertebrates.

Land plants of a kind were present as early as the Cambrian, but it was not until the Devonian that real forests came into existence. In the Carboniferous (often divided into two separate periods by American geologists) the great coal swamp forests were populated by many kinds of amphibians, derived from the crossopterygians; transitional forms are known from the late Devonian. These primitive land vertebrates or tetrapods, which have to return to the water to lay their eggs, were the dominant forms of life of the Carboniferous, so that this period may be styled the Age of Amphibians. That was the beginning of the Tetrapoda, the zoological division including all the four-footed animals and their descendants (snakes, birds, whales and men are also Tetrapoda).

Towards the end of the Carboniferous the first reptiles appeared on the earth. Their mode of reproduction, with a shelled, yolk-rich egg that is laid on dry land, enabled them to advance from the shores and streams and populate the dry uplands. And so the dominance of the continents passed to the reptiles, and the Age of Reptiles began.

It begins with the final period of the Paleozoic, the Permian, and continues throughout the Mesozoic era. In the Permian and Triassic (the first period of the Mesozoic) flourished various primitive reptiles, of which the mammal-like reptiles or therapsids are of special interest; they include the ancestors of the true mammals. In the Jurassic and Cretaceous periods the rule passed to the dinosaurs and some other advanced reptilian forms.

The first mammals appear on the scene around the Triassic-Jurassic transition, so that mammals were in existence through the long reign of the dinosaurs. They remained small, however, and made no bid for dominance of the stage.

At the end of the Cretaceous, the great majority of the ruling reptiles became extinct and their roles were taken over by the mammals. Thus began the Cenozoic era, with which we are concerned here; for this is the Age of Mammals.

The Cenozoic is divided into two periods, the Tertiary and Quaternary (names taken over from an earlier classification which included the Primary and Secondary, now superseded by other categories). The periods, in turn, are subdivided into epochs, of which there are five in the Tertiary and two in the Quaternary. The epochs were originally distinguished by the British geologist Charles Lyell on the relative numbers of modern forms of molluscs found in the fossil marine faunas and named accordingly.

Unfortunately, the result was a string of names that sound confusingly similar. The five Tertiary epochs now recognised begin with the Paleocene ('early dawn of modern life') and continue with the Eocene, Oligocene, Miocene and Pliocene as modern forms of life appear in increasing abundance. In the Quaternary fall two epochs, the Pleistocene or Ice Age, and the Holocene or geological Recent. Obviously this scheme of naming is didactically doubtful; it has been remarked that it is as bad as if a novelist gave his leading characters the names of Tomkins, Tomlins, Tomkinson, etc. The names, however, are entrenched by now, and as we go on I hope to show that each of the epochs has its own distinctive profile.

There have been many attempts to date the geological succession in terms of years, but real advance became possible only with the discovery of radiometric methods. It may be exemplified by the K-Ar or potassium-argon dating method, which is currently the most important. It is based upon the fact that potassium, a very common element in nature, contains a radioactive isotope which disintegrates to form the gas argon (calcium is also produced but not used in dating). The rate of the transformation is known, so that the age of the mineral may be obtained by establishing its potassium/argon ratio. Several other 'radiometric clocks' are also in use. As a result, we now have an absolute chronology of the geological history of the earth, as outlined in Table 1. For our present purpose it is of particular interest that the sequence of Cenozoic epochs and ages (an age is a subdivision of an epoch) has been dated in great detail, mostly on the basis of fossil-bearing strata in North America.

A reader making his first acquaintance with geological dates must be struck by the magnitude of the figures, unless defence budgets and the like have made him immune to large numbers. If so, it is a valuable exercise of the mind to try and visualise the actual meaning of figures like 5,000 million years (the approximate age of the earth) or 65 million years (the duration of the Age of Mammals). Even the yardstick of geological time, the mega-year or year-million, is somehow beyond our imagination. If the tallness of a man is made to represent his longevity, then the mega-year is a distance beyond the horizon. If a human generation is thirty years, the mega-year accommodates 33,000 generations. If the mega-year is compared to an ordinary year, the total of written human history would be less than two days, and a human life three-quarters of an hour. And yet the mega-year is so short that an even greater unit, the giga-year (1,000 mega-years, or a billion years), is sometimes used; the age of the earth is approximately five giga-years. When we look back into the geological past, trying to sense and understand, we seem to stare down into an awesome pit of time where our everyday measures and experience lose all their meaning.

Perhaps, as Milton Hildebrand has suggested, the best way to sense the grandeur of geological history is to relate it to cosmic events. The Milky Way, our galaxy, consisting of some 100 billion stars and an untold number of planets, is rotating slowly. Our solar system lies at some distance from the centre in one of the spiral arms and revolves at its own speed around the hub. It has completed about one revolution since the beginning of the Age of Reptiles. Beyond the Milky Way, other galaxies are found and at least 100 million are within the reach of our telescopes; the most distant are so far away that the light we now receive from them has been on its way ever since the Carboniferous and we now observe them as they looked in the Age of Amphibians. One of the most dramatic events revealed by our telescopes is the collision between two galaxies, which strip each other of interstellar gases as they move through one another, causing a tremendous energy flow. In that segment of the universe visible to us, some 650 collisions have taken place during the Age of Mammals. Each collision lasts about as long as the average life span of a mammalian species.

Table 1. Geological time and the fossil record in outline

Era	Period	Epoch	Some typical animals	Beginning (mill. yrs.)
Ceno-zoic	Quaternary	Holocene	Modern man, domestic animals	0·01
		Pleistocene	Primitive men, mammoth	3
	Tertiary	Pliocene	*Hipparion* (3-toed horse)	12
		Miocene	Bear-dogs, dryopithecine apes	25
		Oligocene	Early apes, primitive mastodons	36
		Eocene	Archaic ungulates & carnivores	55
		Paleocene	Multituberculates, pantodonts	65
Meso-zoic	Cretaceous		Dinosaurs, marsupials, placentals	140
	Jurassic		Dinosaurs, pterodactyls	180
	Triassic		Mammal-like reptiles	230
Paleo-zoic	Permian		Stem reptiles	280
	Carboniferous		Amphibians	345
	Devonian		Fishes	405
	Silurian		Jawless vertebrates	430
	Ordovician		Graptolites, orthoceratites	500
	Cambrian		Trilobites, brachiopods	600
Pre-Cambrian			No good fossil record	ca. 5000

We now know that many year-millions have passed since the fossil we are studying was part of a living being – an animal of the species we call *Hipparion sitifense*. And we can even state that the age probably was between five and eight million years, the middle Pliocene.

Next we have to consider the setting in which our *Hipparion* lived and moved. How do we find out what it was like? What kind of a landscape formed the stage? What kinds of plants covered the earth, what other kinds of animals lived there? Was the climate cold or warm, moist or dry?

The plant and animal life of the time is revealed by the fossil record, which is being studied by an ever-growing array of specialists in different fields. There are specialists on fossil mammals, fossil birds, reptiles and amphibians, on fossil shells, fossil insects, and so on. The paleobotanist works on the plant remains; sometimes the most revealing results are reached by the palynologist, who studies the fossilised pollen grains. From the combined efforts of such workers, a picture of the animal and plant communities of the past is gradually emerging.

Many organisms give good indications of the climate in which they lived and this helps us to chart the climates of the past. This is evident for fossil floras and faunas containing forms that are still in existence; but even entirely extinct forms, especially plants, may tell us a good deal about paleoclimates. For instance, the leaves from the late Eocene of Messel in Germany are large, sharply pointed, with thin cuticles and many openings (stomata), all of this suggesting a warm, moist climate.

An interesting method of direct measurement of temperatures in the past has recently been developed. It is based on the fact that oxygen isotopes are bound into the calcareous tests of marine animals in somewhat varying proportions, depending on the temperature of the water. In this way it has been possible to measure the temperature of the ancient seas.

Now let us return to our Pliocene three-toed horse once more, this time to examine the deposit in which it was found. Its nature will always tell something about the conditions under which it was laid down. Ancient shore lines, for instance, may be indicated by strata of beach sand, or by pebble layers, or by wave-cut benches in the rock. Various kinds of rapidly accumulating sediments called flysches and molasses are formed in the neighbourhood of newly-risen mountain ranges. Fossil sand dunes may suggest aridity and will also show the direction of the prevailing winds. Morainic deposits are laid down by glaciers. Thus study of the sediment itself helps to bring out a picture of the paleogeography of the land. The division into marine and continental deposits, of course, reflects the distribution of land and sea in the past.

Among the many tools which we now have for the study of paleogeography, that of paleomagnetism deserves special mention. Just as a compass needle orients itself parallel to the magnetic field of the earth, so do the magnetic components in the rocks. The difference is that while the compass needle remains free to swing around, the mineral needles are frozen into the rock as a permanent record of the direction of the field at the time when the rock was formed.

Now the remarkable fact is that the direction of the field tends to change as we go to older and older rocks, suggesting that the poles of the earth have moved during geological time. Furthermore, the apparent pole positions recorded in different continents tend to diverge, as if these continents had moved in relation to each other, as well as to the poles. Analysis of these data has now established beyond reasonable doubt that great continental movements have indeed taken place in geological time, and so the old concept of continental drift has finally been vindicated.

In the beginning of the Mesozoic, the land masses of the world were still connected in two great supercontinents. The northern one, which consisted of North America, Greenland, and Eurasia north of the Himalayas, is called Laurasia. The southern continent, or Gondwana, was probably even larger and was later to split up into South America, Africa, Antarctica, Australia, and the Indian peninsula.

The main part of the drift occurred in the Mesozoic and Cenozoic. At the threshold of the Age of Mammals, it had reached a stage where the fragmentation of the land masses was maximal, and it was further accentuated by the fact that large parts of the continental blocks were inundated by shallow shelf seas.

Never before or after has the land area of the world been split up to such an extent. North and South America formed two separate fragments of the original supercontinents. Africa was completely separated from Europe by the Tethys or paleo-Mediterranean and was divided into two or three separate islands by continental inundations. India, Antarctica and Australia, three chips of ancient Gondwana, were radiating out from their original sites. And the great Eurasian land mass was neatly divided into a European and an Asiatic region by a broad interior sea stretching from the Tethys to the Arctic. This gives us eight separate areas, each one a private Noah's Ark with its own fauna and flora, pursuing their insular evolutionary fates.

During the course of the Tertiary, new connexions were being formed to supersede the old ones. In the Paleocene and beginning of the Eocene, a land bridge across the North Atlantic still seems to have been in existence; it foundered in the early Eocene. Meanwhile, a land bridge was formed across the Bering Strait between Siberia and Alaska, and this continued intermittently to permit intermigration throughout the Age of Mammals. The strait between Europe and Asia was bridged for the first time in the Eocene.

The Indian peninsula seems to have become attached to Asia at an early stage in the Tertiary, perhaps in the Eocene. Africa, on the other hand, continued to be almost isolated up to the end of the Oligocene; with the beginning of the Miocene, direct contact between Africa and Eurasia was established. South America remained an island continent even longer, for it was only in the early Pliocene that the land bridge of Central America emerged. And Australia and Antarctica failed to make any contact at all and so remain island continents to this day.

As early as the Eocene, then, the northern continents – the fragments of ancient Laurasia – had become more or less interconnected, so that land animals were able to intermigrate between them; they formed what has been termed the World Continent, a gigantic evolutionary playground over which the terrestrial animals could move more or less freely. In the Miocene, Africa was added to the World Continent; in the Pliocene, South America.

We have considered some of the methods that are used to reconstruct the past. Let us now proceed to the main actors on our stage, the mammals.

The Age of Mammals lasted hardly more than one-third of the total time that mammals have existed. Primitive mammals were in existence as early as the beginning of the Jurassic, 180 million years ago, and probably even earlier. This means that the mammals lived for more than a hundred million years in a world dominated by reptilian forms.

Zoological classification makes a clear-cut distinction between mammals and non-mammals; the former are included in the class Mammalia, the latter are not. But the mammals arose by gradual evolution from reptiles and there are no real boundaries in the sequence; there is no point at which we can say that the reptile suddenly became a mammal. Rather there are one-quarter mammals, half-mammals, and so on. These transitional forms, representing several different populations becoming more and more mammal-like, are found in the late Triassic and the early Jurassic.

An osteological difference between reptiles and mammals, which is at present used to draw a somewhat arbitrary dividing line between the two classes, is found in the joint between the skull and the lower jaw. The reptilian lower jaw is constructed of several bones, and the joint between it and the skull is formed by the articular bone (in the lower jaw) and the quadrate (in the skull). Typical mammals have only one bone (the dentary) in the lower jaw and the joint lies between it and a skull bone called the squamosal. The articular and quadrate, however, have not vanished, but have moved into the inner ear to form two auditory ossicles, the malleus and incus. Such a transition seems almost incredible at first sight, but the fossil record shows how it took place. In the transitional forms, the joint in the skull was formed by both the squamosal and the quadrate, while the corresponding joint on the jaw included both the dentary and the articular, so that both the mammalian and reptilian joints were functioning simultaneously. The articular and quadrate were close to the ear region and became free to take part in sound transmission as soon as they lost their function in the jaw joint.

Unfortunately, our knowledge of Mesozoic mammals is rather inadequate. Most of these creatures were very small and their minute teeth, skulls and bones are hard to find. New methods, however, in particular the systematic washing and sieving of large amounts of matrix, are now producing a hitherto unexpected wealth of material. Another very promising line of attack is the systematic searching for ancient cave and fissure fillings, which often contain great amounts of bones of small animals.

The mammal-like reptiles or therapsids of the Permian and Triassic give us an idea of the reptilian ancestry of the mammals. They approach mammals in many other ways besides the jaw joint. Their dentition shows the beginning of a differentiation into front teeth or incisors for nipping and grasping, enlarged eye-teeth or canines for piercing, attack or defence, and expanded cheek teeth for chewing. This suggests a need for rapid food utilisation, which makes us suspect the therapsids were active beings, perhaps warm-blooded or nearly so. A modern mammal keeps a nearly constant body temperature and hence is able to function in a cold climate where a reptile would fall into torpor. Possibly the therapsids were approaching this condition.

This is also suggested by the construction of their palates (roof to the mouth), which reroutes the air tubes around the mouth cavity, so that breathing can go on undisturbed by chewing. Their rib cage tended to become restricted to the fore part of the trunk, suggesting the presence of a muscular diaphragm – the septum that separates the breast cavity from the belly and makes an efficient breathing mechanism. In the nasal cavity there may be turbinal bones, as in the mammals; covered with mucous tissue, they serve to clean, moisten and perhaps warm the air before it enters the lungs, as well as being used for smelling. All this may suggest metabolic processes at a much higher rate than in normal reptiles.

The tritylodonts, a group of mammal-like reptiles in the late Triassic and early Jurassic, were so close to the mammals in structure that their position is disputed. They are also interesting in being the first of a long series of mammalian forms evolving a rodent-like skull with large, gnawing front teeth, separated from the cheek teeth by a long toothless gap or diastema.

In the Jurassic we meet with several distinct groups of mammals; some of them have recently been shown to have precursors in the late Triassic. The triconodonts, some of which were as large as a cat, are thought to have been carnivorous in habits, while some other groups of primitive mammals may have been insectivorous; all of these, however, became extinct in the Jurassic or early Cretaceous, leaving no descendants, unless possibly the living Australian monotremes (the platypus and the spiny anteater) are related to them.

The herbivorous element is represented by the Multituberculata, which arose in the Jurassic and persisted well into the Age of Mammals. They took over where the tritylodonts left off and were the pseudo-rodents of their time, with enlarged, chisel-like incisors; yet they were not related to the tritylodonts, but evolved these characters independently.

Finally, there is a group called Pantotheria, destined to give rise to nearly all the mammals of the present day. They were small forms, probably in the main insectivorous in habit and perhaps not unlike the shrews of today. In the Cretaceous this group gave rise to two separate lines of descendants: the marsupials or pouched mammals, and the placental or higher mammals.

The marsupials, in the guise of small opossum-like creatures, were apparently very common in the later Cretaceous. It may be assumed that the female already had a pouch on the belly, in which the young were reared after birth. In this way the young have a sheltered environment for development, though the actual gestation period is kept at a minimum. Osteologically the marsupials can usually be recognised by a peculiar inflection of the angle of the jaw and by the marsupial bones, if preserved.

The placental mammals have a placenta, which supplies oxygen and nutriment to the embryo from the mother's blood. The gestation period is thus extended and the young are born at a more advanced stage of development. The earliest placental mammals may well be regarded as members of the order Insectivora in a broad sense, but in the late Cretaceous there is also a tendency to diversification, and several different orders of placental mammals appear to have evolved well before the beginning of the Paleocene. Evidently they originated from the earliest, insectivore-like placentals; our own ancestry, too, must be located somewhere among these little, quick-moving creatures from the last lap of the Age of Reptiles.

We are now at the very threshold of the Age of Mammals, in late Cretaceous times, and we may as well look about, for this will be our last glimpse of that peculiar scene. The time is some sixty-five or sixty-six million years ago. Dinosaurs are still much in evidence, both in the lush rain forest and the tropical swamps. The drier country is populated by large herds of the great horned dinosaurs; they are stalked by *Tyrannosaurus*, largest and most terrifying of predators. In the swamps may be seen the long-necked, long-tailed kind of dinosaur, while the duck-billed dinosaurs browse in the forest. Other, smaller reptiles roam the shores and woods, some of them probably an active menace to the mammals – *Tyrannosaurus* probably was too big even to notice a contemporary mammal, especially one that was playing 'possum.

In the undergrowth and trees, the mammals are going about their own business. The rodent-like multituberculates are foraging for nuts, roots and the like, while the marsupials and insectivores are feeding on grubs, insects and eggs (including those of the dinosaurs). All keep a wary eye for the small carnivorous dinosaurs, the big python-like snakes and other dangers. Their own next of kin may be dangerous too: there are some small, undifferentiated ungulates, primitive hoofed mammals that were almost certainly omnivorous and ready to turn into meat-eaters given the chance.

But perhaps to us the greatest thrill of all is to recognise, in this strange setting, some very early members of our own zoological order, the Primates. They have only recently been identified but this shows that there remains much to be learned about the late Cretaceous mammalian fauna.

Birds are plentiful and have perfected the art of flying since the Jurassic when the first, primitive bird fluttered and glided from tree to tree. Now there are loon-like and stork-like birds about, also several kinds of wading birds, as well as a few primitive, toothed forms. The flying lizards, of remarkable and hideous appearance, are already gone.

The seas still teem with predaceous reptiles of monstrous sizes and shapes, but here too the reptilian reign is nearing its end. We are now moving towards a curtain fall and a total turnover without parallel in geologic history. Why these balanced, varied communities should be so suddenly disrupted at the end of the Cretaceous, we do not yet know. Many theories have been advanced to account for the mass extinction at this time, but none is entirely successful. One of the problems is that the details of the story are very hard to come by. Locked in our own time, we look at these events over a chasm of 65 million years. The distance blurs the picture; intervals of up to 100,000 years are hard to assess. It may be that there was a gradual thinning out of dinosaurs at the end of the Cretaceous, but did the final blow fall at the same time all over the world? We cannot tell. And yet there comes a definite level when it is clear that the giants have gone to rest and we enter a world seemingly built to a smaller scale. The Age of Mammals had begun. In time it was to produce giants to dwarf all but the largest of the dinosaurs; it was to produce a more wondrous diversity, more remarkable specialisations, more activity, efficiency and intelligence than any previous Age.

But when the curtain was rung down for the Age of Reptiles that was still in the distant future.

Of course, many kinds of vertebrates – both fishes and tetrapods – survived the Cretaceous/Tertiary transition. The survivors include the mammalian groups of the late Cretaceous: the multituberculates, the marsupials and the placentals. They also include numerous reptiles: the lizards and snakes, crocodiles and turtles, and even some primitive reptiles now represented by a single relict form, the tuatara. Again, the birds crossed the threshold and indeed entered upon an evolution and diversification of such magnitude that the Cenozoic might well be called the Age of Birds. Frogs and salamanders have also come down to us across that great frontier in time; and among the fishes, the main changeover – the expansion of the modern bony fishes – had already occurred in the middle Cretaceous.

The number of those that fell by the way is tremendous. Many of the typical invertebrates of the Mesozoic seas became extinct. All of the dinosaurs, large and small alike, vanished completely – or almost completely – and the same fate befell all the great sea-going reptiles.

Whatever the reason, the outcome was an impoverished world, presently to be conquered by the mammals. At this time, it should be noted that some ways of life, or 'ecological niches' to use the scientific term, had already been occupied by the mammals for many million years. The multituberculates had established themselves in the 'rodent niche' and continued successfully in charge. And the 'insectivorous niche' was also occupied by mammals since time immemorial. Now the niches once held by the dinosaurs were vacant; most were soon to be filled by mammals.

The niches tend to persist, their occupants to change. It is as if there were a standing set of roles in a play, while the actors filling them are forever changing. There is the primary division into plant-eaters and meat-eaters. Among the former, within the 'gnawing' niche, we have already seen that the tritylodonts cede their place to the multituberculates; and the latter are in time going to be supplanted by the true rodents. There are many other distinct roles among the plant-eaters, for instance the very large forms that rely on sheer size and strength for protection: the elephants and hippopotami of the present day, the giant dinosaurs of the Mesozoic. Other large herbivores may carry defensive horns: the horned dinosaur in the Cretaceous, or today's rhinoceros. Then there is a roster of medium-sized herbivores whose main defence is rapid flight – the horses and antelopes of the modern fauna, certain bipedal dinosaurs in the Mesozoic. Some plant-eaters encase themselves in a coat of mail like the armadillos of the present day, or the armoured dinosaurs of the Age of Reptiles.

The flesh-eaters are less divergent, yet they also have their standing roles; as exemplified in the modern fauna, these include the predaceous stalkers (cats and foxes), the pursuers (wolves and cheetahs), the omnivores (bears), the scavengers (hyenas); there are also the insect-eaters (shrews).

One of the more surprising results of paleontology is the discovery that the roles changed so often and so radically during the Age of Mammals. To be sure they are hereditary to a degree; there are, so to speak, dynasties that keep their special role for millions of years. But as a rule the dynasties come to an end and an entirely new family of actors step into the same role. For instance, the mammals performing the role of giant herbivores form a succession of unrelated groups, not a single lineage. The roles persist, unchanging like Columbine, Pantaloon and Harlequin; but while the current performers go through their motions on the stage, the next set may already be waiting in the wings. Now why should this be so?

Throughout the history of the organic world, there has been a dominant pattern of evolution that very aptly was termed 'adaptive radiation' by the American paleontologist H. F. Osborn. This is the evolution of diversified descendants out of a single ancestral stock and in this way a continuous supply of new actors is created, ready to step into the old roles. For the great play is enacted on innumerable stages all over the world; and although the role, or niche, may be monopolised by a certain group of mammals in most parts of the world, there may still be some area where that group has failed to penetrate. Here the niche is empty and there is a chance for local talent to step into it. Thus a competing form may evolve locally, perhaps later on to spread to other areas where it can crowd out and replace the previous players.

Now we can see why the excessive fragmentation of the land world at the beginning of the Age of Mammals was so important. Each local Noah's Ark produced its own adaptive radiation. When finally the arks drifted into contact, a competition ensued, with natural selection as judge and jury.

Very often, superiority results from some advantage in general efficiency. For instance, the herbivorous giants that arose in the early Tertiary had a primitive dentition, poorly equipped to cope with the great amounts of green fodder that they required. While these giants lumbered on their way, keeping to the most succulent herbage so as not to wear out their teeth prematurely, other plant-eaters evolved more efficient teeth. In due time they became so superior in getting and utilising food that they were able to grow to great size and oust the archaic beasts. The living terrestrial giants, the elephants, indeed have one of the most efficient sets of teeth that ever evolved.

There are also other ways in which efficiency may be increased, enabling new groups to surge to dominance. A perfection of the brain and nervous system, and the sense organs, is obvious in the fossil record; so is a gradual mechanical perfection of the skeleton and (presumably) musculature. No doubt the physiological efficiency was improved in many other ways as well – for instance the intestinal apparatus, the ductless glands, the placenta – though this is difficult to read out of the fossil bones. And so the overall picture of evolution in the mammals shows not only radiation into an incredible number of adaptive types, large and small, but also the gradual perfection of the general level of efficiency within most evolving lines.

Before we are ready to go on to our main story, some final remarks on names. In order to talk about animals, we must be able to refer to them exactly. Most of the extinct mammals have no vernacular names in any language, so we have to use their scientific ones. This has some drawbacks (the names may be long and difficult) but also has the great advantage of never being equivocal: a scientific name means just one kind of animal. In contrast, the vernacular 'elk' means *Alces alces* in Europe and *Cervus elaphus* in America; 'panther' in India means the leopard, but 'panther' in Florida means the puma; 'tiger' in South America means the jaguar, and so on.

The basic unit in classification is the species. Volumes have been written about what a species is, and this is not the place to review them. *Felis silvestris*, the cat, is a species, exemplified by the domestic cat but also by the wild cat, distributed in Eurasia and Africa, from which the domestic form has been derived. Close relatives to the cat belong to the same genus, *Felis*, for instance the lynx (*Felis lynx*) and the puma (*Felis concolor*). Slightly more distant relatives are put in different genera, for instance the tiger (*Panthera tigris*), but belong to the same family, the Felidae or cat-like carnivores. (Genus and species names are always written in italics, names of families and higher categories are not.)

Other carnivore families are the Canidae or dog-like carnivores, and the Ursidae or bear-like carnivores. They all belong to the order Carnivora; while the horse, for instance, belongs to the order Perissodactyla, and the elephant to the order Proboscidea. Finally, they are all members of the class Mammalia, or mammals. In this way the classification attempts to formalise the evolutionary relationships of the various species of mammals, and they are grouped into orders, families and genera to reflect our view of their affinities.

So extremely varied are the tetrapods (mammals, birds, reptiles and amphibians) of the Age of Mammals that the reader may be in danger of getting lost in a jungle of names. For quick reference, a classification of all the tetrapod orders is given as an Appendix (p. 227).

Chapter 2

The Paleocene: epoch of conquest

> I feel dizzy, standing on this eminence
> and looking down the long ages which have passed
> like the waves in the Sound, and which have left
> such almost worn-out vestiges of the past world;
> and which can now speak in a whisper only, when
> everything else has become silent.
>
> Carolus Linnaeus: *Scanian Voyage,* 1749.

THE AGE of Mammals begins with the Paleocene, in some ways the most interesting epoch of the Tertiary, but also the least known. Now at last the mammals, free to expand into the ecological niches vacated by the extinct reptiles, entered upon their great radiation and evolution. An extraordinary amount of evolutionary change seems to have taken place in the Paleocene, for at the end of the epoch the earth was already populated by a varied, balanced mammalian fauna. Yet its record is incomplete and often confusing, as if comprising only a few pages torn at random from the opening chapter of a book.

Recent work in potassium-argon dating suggests one reason why this should be so. In spite of the great amount of evolution that took place among the mammals, the Paleocene was a comparatively short epoch, only about half the length of the succeeding Eocene epoch. It follows that the amount of fossiliferous deposits formed in the Paleocene was smaller than that of the Eocene. Of course, the deposits laid down in the Paleocene have also been longer exposed to erosion than those of later date, so that their rate of loss is likely to be greater.

There are also reasons to suspect that Paleocene mammal localities are harder to discover than others. It takes a specially trained eye to observe the teeth or bones of small mammals like those of the Paleocene. These creatures were obscure in life and they remain so in death. But modern prospecting and collecting methods will probably revolutionise our knowledge of the beginning of the Age of Mammals in years to come.

The climate of the Paleocene appears to have been warm and humid in general. Paleotemperatures indicate that the seas have cooled gradually throughout the Tertiary, probably from a maximum some time in the Cretaceous. In Australia somewhat conflicting results have been obtained, for here the paleotemperatures of the early Tertiary (about 10–12° C) are distinctly lower than those for the Oligocene and Miocene (16–20° C). The

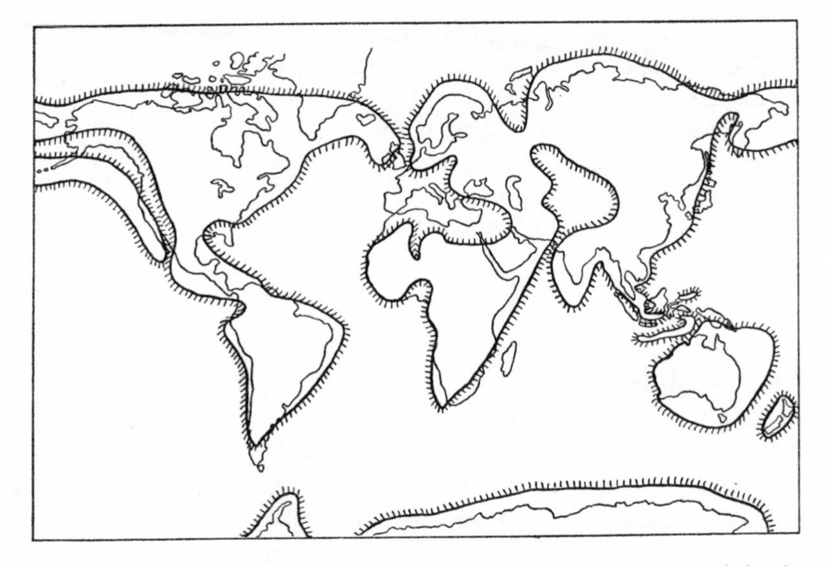

Figure 3. Approximate distribution of land and sea at the beginning of the Age of Mammals.

solution to this seeming discrepancy may lie in the movements of Australia relative to the South Pole. In the early Tertiary, Australia was in a position extending from 30° to 60° south or some twenty degrees nearer the pole than now. The drifting towards warmer latitudes probably overshadowed the effects of the gradual lowering of the temperature in the earlier Tertiary.

With this exception, most of the evidence suggests that the climate was much warmer than today. For instance, rich fossil forest floras are known from areas that now belong to the most forbidding part of the Arctic: not only in Iceland, but also in Greenland, Spitzbergen and Grinnell Land, which, at 81° 45′ N, has the northernmost fossil flora of the Tertiary. The exact date of these forests is not yet known but they are definitely from the early part of the Tertiary. Floras definitely datable as Paleocene are unfortunately rare, but the few known corroborate the impression that Paleocene climatic conditions were essentially the same as in the Cretaceous. On the other hand, there is a suggestion of a temporary cooling during part of the Paleocene in Europe, but it is not yet clear whether this is a general or a local phenomenon.

One of the main problems of the Paleocene is to find out about the crews that manned the arks on the various continental fragments produced by the breaking-up of Laurasia and Gondwana. So far we have records of Paleocene mammals only from North and South America, Europe and Asia, and it is only in North America that a reasonably detailed succession is known. For Africa, India, Australia and Antarctica we have no records at all of Paleocene mammals; and so our story must begin in North America.

In the late Cretaceous, the waters of the present area of the Gulf had extended in a great arm to the north and northwest, into the southern parts of Alberta, Saskatchewan, and Manitoba; but it contracted gradually, so that by the end of the Paleocene only a belt some 100–200 miles wide along the present-day Gulf Coast, including Florida, was inundated. As a remnant of the old inland sea, a bay stretched northward along what is now the Mississippi, as far as the mouth of the Ohio; this bay was to remain a prominent geographic feature throughout the Paleocene and Eocene. Parts of the east and west coasts were also inundated, but the main part of North America was dry land throughout the Age of Mammals.

Mammal-bearing Paleocene strata have been found in only a few places in the world; the most important sequence comes from the Rocky Mountains, from New Mexico in the south to Wyoming and Montana in the north. The Rockies were originally brought into being by the so-called Laramide revolution, a cycle of mountain building beginning in the late Jurassic or early Cretaceous, with a climax in the late Cretaceous. The folding and uplift continued well into the Eocene and possibly even the Oligocene. Thus in Paleocene times the area was bristling with young, rugged mountain ranges, with plenty of active volcanoes.

The essential structures of this early version of the Rockies were the same as those of the present day. The separate mountain arches enclosed great intermontane basins; both the arches and the basins are still there to be seen. Here was a varied environment for the unfolding mammalian life, ranging from jungles, swamps and lakes in the valleys to a cool-temperate habitat in the high mountains.

Warm conditions in general are suggested for North America; as an example, it has been found that the 20° C annual isotherm on the Pacific coast lay north of 49° N in the Paleocene, while its present-day position is more than twenty-two degrees further south. An analogous shift today would give Denver, Colorado, the climate of Mexico City. Since the present-day high level of the Rockies is due to relatively recent uplift, the intermontane valleys were still lowlands in the early Tertiary and received abundant rainfall, except perhaps in the lee of some mountains. As a result, they were clad by dense forests of a sub-tropical type, except for the deeper depressions where bogs or lakes were forming; in the bogs, the rank vegetation accumulated, finally to form beds of lignite or brown coal. In the forests were both subtropical forms like the fig, banana, sabal palm and soapberry, and temperate forms like the oak, sycamore, redwood and ginkgo.

As time passed, the basins were gradually filled by deposits produced by erosion of the enclosing mountain ranges. By late Oligocene times the area had been converted into a high-lying plain. The original mountain ranges were still there, but they were now buried up to their necks in their own débris; only a few remnants were visible, rising as so-called monadnocks over the even surface. Deep down in this enormous accumulation of strata rested the record of early Tertiary life.

Rejuvenation of the mountains began as uplift was resumed in Miocene and Pliocene times. The old basins were reopened by downcutting rivers. Gradually the record of past life was dissected out of the old strata; and once again, as the débris was cleared away, the ancient mountain world began to emerge.

The southernmost of the basins is the San Juan Basin in New Mexico, which lies partly in the San Juan River valley, partly across the continental divide. To the west flows the San Juan to join the Colorado on its long way to the Gulf of California; to the southeast is the Rio Grande, embarking on a still longer journey to the Mexican Gulf. The Paleocene deposits in the basin rest on top of late Cretaceous sandstones; they consist of clays and sandstones with intercalated beds of lignite – the coal-like stuff produced by the decaying vegetation in the ancient bogs. They can be divided into two successive units with quite distinct faunas: the Puerco, which is the oldest, and the Torrejon. An even younger Paleocene fauna occurs in the Tiffany beds in the northern part of the basin.

An even longer series of Paleocene mammal-bearing deposits is found in the Bighorn Basin in northern Wyoming. Again the Paleocene strata, reaching a thickness of some 3,500 feet, rest upon late Cretaceous dinosaur-bearing deposits. But here there are four successive faunal levels; in addition to the Puerco, Torrejon and Tiffany stages there is a fourth stage, dating from the very end of the Paleocene and called the Clarkfork.

The picture is filled out by sequences in other basins such as the Green River and Wasatch Basins of southwestern Wyoming, the Uinta Basin straddling the Utah-Colorado state line, and the Crazy Mountain Field in central Montana. Sometimes a fifth stage, the Dragonian, intermediate in age between the Puercan and Torrejonian, is recognised; it is represented by a fossiliferous deposit in the Dragon Canyon of central Utah.

Besides the great fossil fields of the Rocky Mountains there are a few scattered sites, for instance in California and Canada.

The Puerco fauna is of extraordinary interest in giving us the first glimpse of vertebrate life just after the extinction of the dinosaurs. The environment with its great jungles, swamps, lakes and mountains, had probably changed very little since the Cretaceous. But the animal world appears wretchedly impoverished, when compared with the variety of impressive creatures that populated this same area in the recent geological past.

Such large animals as were still in existence are to be found among the reptiles of the lakes, rivers and swamps. Here there were both true crocodiles and alligator-like forms, and also the peculiar, big lizard called *Champsosaurus*, a long-snouted form the size of an alligator. Like the crocodilians, this animal dated back to the age of the dinosaurs, but it was not related to them. The champsosaurs belong to an ancient reptile group, allied to the living tuatara of New Zealand.

The rivers and swamps must have teemed with smaller forms of life. Fishes include sturgeon, garpike (*Lepisosteus*) and bowfin (*Amia*). There were many kinds of turtles, some of them very archaic forms belonging to the extinct amphichelydians, in which the head could not be drawn into the shell; others were related to modern swamp turtles and leathery turtles. Land tortoises are not yet found, nor are the snapping turtles.

On land, too, smaller reptiles must have been abundant, although the record at hand is incomplete and the presence of some can only be inferred from fossils in earlier and later strata. They were related to modern skinks, iguanas, slow-worms and monitors. The snakes were big boa- or python-like forms; no poisonous snakes are known.

There were no mammals to rival the great swimming reptiles; the maximum size of the Puercan mammals was that of a large dog and most were much smaller. Most of these beasts resemble each other rather closely, but there is one group of mammals which diverges sharply from all the others. It would probably have caused a lay onlooker more wonder and amazement than all the other forms together. We have met them already: the multituberculates, which unlike the other Paleocene mammals belonged to a very ancient order, nearing the end of its long life span.

Early forms of multituberculates are known well down in the Jurassic; the group had been abundant and successful since late Jurassic times. By now, in the early Paleocene, they had existed for rather more than a hundred million years and were still going strong. When these hardy old-timers finally passed out of the picture, many million years later, they had set a record for longevity of mammalian orders which has not been broken yet.

The multituberculates had large, chisel-like incisor teeth, like rodents. The molar teeth are elongate, with parallel fore-and-aft rows of cusps,

Figure 4. A member of an extinct branch of primitive mammals, the multituberculate *Ptilodus* measured slightly below a foot (30 cm.) from nose to base of tail. Ptilodonts of this general type were common in the Paleocene forests of North America and Europe.

forming an efficient chopping device. In some forms, like *Ptilodus*, the lower hind premolar was much enlarged, forming a big, striated cutting blade, perhaps for opening husked fruits. Most ptilodonts were comparable to squirrels in size; they range in time from the Cretaceous to the Eocene. Skeletal finds indicate that the normal stance was a fairly primitive one with strongly flexed limbs and the foot stood flat on the ground (plantigrade). The 'big' or inner toe was somewhat opposable (actually the middle toe was the largest and the others decrease symmetrically around it). The foot thus had some ability of hand-like grasping, like a monkey's foot but much less developed. It may suggest that the ptilodonts were climbing,

Figure 5. Largest of the multituberculates were *Taeniolabis* and its relatives, of North America and Asia; this group is not found in Europe. They were sturdily built, somewhat marmot-like creatures.

tree-living forms. Very outlandish and semi-reptilian would these small creatures have looked to us.

A second group of multituberculates, which appeared in the late Cretaceous and persisted in the Paleocene, is formed by *Taeniolabis* and its relatives. They were broad-skulled animals about the size of a terrier, with powerful gnawing incisors. They lacked the big premolar blade of the ptilodonts and presumably led a quite different kind of life. Apparently too large to have been good climbers, the taeniolabiids may have been ground-living or possibly swimming forms.

Apart from some opossum-like marsupials, the remaining Puerco mammals are placentals. They were still rather close to their common insectivore origin and it might almost be justifiable to regard them as constituting a single order of mammals. It is only because their descendants became so much more varied that we try to identify different orders in the early Paleocene. We classify the Puerco mammals not according to what they were themselves, but because of what their descendants would become in later times.

They were small or medium-sized. They tended to be fairly long of body, with relatively short legs and a long, tapering tail. Their fore-paws and hind feet were five-fingered and five-toed and carried claws – blunter and more hoof-like in some forms, sharp in others. Most of them still moved about with the heels and palms touching the ground at each stride: their gait was plantigrade, like that of a bear. The head was long, with a powerful jaw apparatus carrying a complete, primitive set of teeth, usually with well developed eye-teeth or canines. The brain was small and compared to a modern mammal like a horse, dog or cat, any one of these archaic beasts would have appeared very dull.

Across this fairly uniform group there is an incipient division into plant-eaters and meat-eaters, with some suggestions of the beginnings of specialisation. Some of the most specialised forms at this date were the taeniodonts. One of these, *Wortmania,* had reached the size of a large dog but was more massively built, with great canine tusks and blunt-cusped cheek teeth; its feet carried large claws, perhaps for digging up roots and bulbs. It was the most powerful mammal of its day, a miniature blueprint of the giants of times to come. Here were the beginnings of one of the distinctive early orders of mammals.

The primitive hoofed mammals or Condylarthra must also have been much in evidence. They came in various sizes, most of them small but a few, like *Ectoconus,* as large as *Wortmania.* A complete skeleton of this animal has been found in the lower levels of the Puerco. This individual did not lead an enviable life, for its right elbow joint was severely diseased and the animal must have been lame. *Ectoconus* stood about 1½ feet high and measured 3 feet from the nose to the beginning of the tail; the tail was over 2 feet long. With its long, flexible body and short, rather heavy limbs, it looks almost like a diagram of an ancestral, generalised ungulate.

Numerous smaller forms share the tendency to develop somewhat blunted cusps on the cheek teeth, suggesting that they fed mainly on vegetable matter. Yet the entire assemblage gives the impression of overgrown insectivores with a vegetarian bias.

In another group the cheek teeth tended to retain somewhat sharper cusps, and the canines are long, piercing fangs. These are the arctocyonids, sometimes regarded as early carnivores, sometimes as Condylarthra. The arctocyonids, though not related to the true bears, probably resembled them to some extent in their habits and had a mixed diet. The arctocyonid family, which arose in the late Cretaceous, is represented in the Puerco by animals about the size of a fox; and the skulls, with their long jaws and sharp teeth, do have a carnivorous look about them, which is somehow lacking in other condylarths.

Of still greater interest to us are the earliest known members of our own mammalian order, the Primates, which have recently been identified both in the late Cretaceous and early Paleocene of Montana. These animals, which have received the picturesque generic name *Purgatorius* (after the locality, Purgatory Hill in eastern Montana), are unfortunately known only on the basis of some isolated teeth. The characteristics of the teeth indicate rather close relationships to insectivores and condylarths, and so suggest that the Primates were still rather close to the point in time when they became differentiated from the basal placental stock.

Finally, a number of insectivores swell the ranks of the mammalian fauna such as we see it on the threshold of the Age of Mammals. It is true that the Puercan does not give us a complete picture of the early fauna; it is as if we tried to visualise the modern mammals of the world on the basis of, say, fifty species collected at random in a restricted area. Some forms are obviously missing, for instance the lemur-like basal Primates from which monkeys, apes, and man are derived. But even if future discoveries fill in many blanks, probably they are not going to affect the main impression of the fauna, which is one of actors in rehearsal, not yet in dress and make-up, falteringly trying out their roles – awkward understudies trying to revive the play left unfinished by a defunct company. Here is a world where everything is groping towards future opportunities.

The story of how the mammals grew into their new roles is told during the remainder of the Paleocene. By Torrejonian times the roster of known forms has swelled greatly, trends that began in the Puercan have continued, others are being initiated, and the range of size and specialisation is continually growing broader.

The multituberculates continue to flourish. The insectivores are present in astonishing variety and one gets the impression that these animals swarmed on the ground, in the trees, along the shores. Many of the Torrejon insectivores look like grand-uncles of various later forms. One animal seems to be an early relative of the present-day tree shrews of southeastern Asia; another group seems transitional between insectivores and carnivores; and yet another appears to be closely related to the ungulates. But this last-mentioned group, the Pantolestidae, set out on a quite different tack and became adapted to an amphibious existence, somewhat like the living desmans.

Besides other conservative insectivores, at least two groups evolved chisel-like incisors and established themselves as pseudo-rodents – always, it seems, a lucrative niche.

Several forms of Primates are now present, but those known to us in the middle Paleocene still look like offshoots, away from the mainstream of later primate evolution. Oddly, a trend towards a pseudo-rodent condition seems to be almost universal among these forms.

The best-known family is that of the Plesiadapidae; a complete skull and most of the skeleton of the genus *Plesiadapis* have been found (the skull comes from Cernay in France and is slightly later in date than the Torrejonian). This is definitely a pseudo-rodent with large chisel-like incisors and a long diastema (toothless interval) between them and the cheek teeth. It is now thought possible that the true rodents may have evolved from plesiadapid Primates. *Plesiadapis* was about as big as a beaver and had well-developed claws, not nails.

Another family, the Paromomyidae, continues the line which we have already met in the Puerco *Purgatorius*. The third family, the Carpolestidae, is perhaps the most astonishing of them all, for they emulated the

Figure 6. A small, rodent-like primate, *Plesiadapis* of the Paleocene in North America and Europe is now thought to be close to the ancestry of true rodents. It was about the size of a squirrel.

multituberculates even so far as to develop greatly enlarged premolar shears, operating against many-cusped upper teeth, very much like the *Ptilodus* group. This family is now often referred to the Insectivora. All this competition ought to have made it hot for the multituberculates themselves, for they were of course an ancient model lacking many of the refinements of these new rivals. But they withstood it well, and reached the end of the Paleocene without any notable decrease in abundance or variety.

The taeniodonts are still on the scene and are now evolving into larger and heavier forms like *Psittacotherium*. With a head about ten inches long, this animal was the size of a large dog, but extremely heavy and massive in build, with deep, powerful jaws. Presumably it lived on leaves and fruits and it had large, strong eye-teeth which were heavily worn and grew throughout the life of the animal, like the gnawing teeth of rodents.

The primitive ungulates or Condylarthra were still becoming more numerous and varied. There are also members of a new mammalian order, the Pantodonta. Destined to become in later year-millions the giants of their world, they were as yet of moderate size: *Pantolambda* reached a length of some five feet, not counting the long tail, so that it was about as big as a leopard. Of course, being an ungulate, it was herbivorous in habits. Its canine teeth could probably be used in defence, but otherwise it did not possess any protection apart from sheer size, which made it the giant of its time.

Actually the enemies at hand do not appear very terrifying, as long as the pantodonts kept away from the crocodiles and alligators of the streams. We have already met the arctocyonids, the somewhat bear-like condylarths which were probably part-time flesh-eating forms. Another group of omnivorous condylarths, the Mesonychidae, was later to produce some monstrous types, but in the Torrejonian they were still of moderate size; *Dissacus* was about as large as a coyote.

True carnivores are also present in the Torrejon; they belong to a basal, undifferentiated family, the Miacidae, with small, mainly forest-living forms. The Carnivora differ from the various pseudo-carnivores by the presence of enlarged, shearing carnassial cheek teeth; in the upper jaw this function is filled by the hindmost premolar; in the lower, by the foremost molar.

In the upper Paleocene, the Tiffany and Clarkfork, still more is added to the fauna. The pantodonts which started with *Pantolambda* are now in full flower and the dominant mammals of their time. Their largest members, *Titanoides* and *Barylambda,* attained about the length of a cow, though they stood a little lower, with heavy, five-toed limbs; probably those heavy paws could deliver a blow to stun any one of the carnivores of the time, however thick-skulled. One of the most curious features of *Barylambda* was its large, powerful tail, almost dinosaurian in appearance; it differed entirely from the cat-like tail of *Pantolambda* and the reduced appendage of *Titanoides.* (See plate 1.)

Insectivores and multituberculates flourished and the Condylarthra continued to produce the main herds of hoofed animals, threading their way through the lush palm groves of the Rocky Mountain basins. But there are innovations too, some of them of great potential importance. Recently, for instance, an eohippus – a member of the order Perissodactyla, or odd-toed ungulates – has been identified in late Paleocene strata in Wyoming and Baja California.

At a locality called Bear Creek in Montana, a peculiar and aberrant assemblage of this date is turning up. The place is now a coal mine, but in the Tiffanian it was a swamp, probably along the edge of a lake. Here a peculiar little mammal would have been seen gliding through the air between the cypress stems, supported by skin folds stretched taut between the arms and legs. It resembled the present-day colugo or 'flying lemur' of southeast Asia. Whether the Paleocene *Planetetherium* was in fact directly related to these specialised insectivores, or whether the similarity was produced by convergent adaptation, is not certain.

The coal swamp was evidently forbidden ground to the small terrestrial animals; only the large taeniodonts and condylarths dared splash into the water. Normally the coal swamp forest belonged to gliding *Planetetherium* and tree-living Primates and insectivores.

Among the arboreal host, however, is found a remarkable newcomer, a true rodent; this is a somewhat squirrel-like animal called *Paramys,* the earliest member of the order Rodentia. The appearance of this newcomer in such an unusual setting is at first sight puzzling but should be connected with the extreme rarity of multituberculates. Evidently the swamp represented an environment which was difficult to penetrate for the otherwise ubiquitous ptilodonts. Perhaps, with their awkward limbs, they were unable to move from tree to tree in the swamp cypress woods, where distances were greater than in the dense jungle that was their normal habitat. However this may be, multituberculates were rare at Bear Creek and so there was an opening for the earliest rodents to exploit. Apparently they were not ready yet to compete directly with the multituberculates. Later on they would do so and finally crowd out their archaic predecessors. The multituberculates, which were seemingly never bothered by the competition offered by rodent-like primates and insectivores, met their match in the Rodentia. But this was still in the distant future when the first semi-squirrels established themselves at the lakeside of Bear Creek.

Another interesting newcomer in the late Paleocene is a member of the order Edentata, represented today by the sloths, anteaters and armadillos. Their forerunners in North America, which are called palaeanodonts, ranged in time from the Paleocene to the Oligocene and were small, short-legged and long-tailed animals with big claws and a long snout. These animals, in which the insectivore ancestry is still plainly to be seen, never amounted to much in North America. But some members of the clan got over to South America, in the Paleocene or perhaps even the Cretaceous, and were soon to deploy as a mighty host.

A number of big, rodent-like animals represented an 'orphan order', the Tillodontia, which culminated in a bear-sized form in the Eocene, then died out. A more important group is that of the uintatheres or Dinocerata, the order that was to take over from the pantodonts as giant plant-eaters. They already rivalled the large pantodonts in size and were much more remarkable in appearance; but their main story comes in the Eocene.

Finally, the late Paleocene saw the advent of the Creodonta – an order of carnivorous mammals in which the carnassial teeth were not formed by the upper fourth premolar and the lower first molar, but by teeth further back in the jaws. The late Paleocene forms were about the size of a glutton and may well have resembled that intrepid hunter in habits.

Such, briefly reviewed, was the opening phase of the Age of Mammals in North America. Let us now turn to what is known about the Paleocene history of the Old World.

While most of the North American continent was dry land throughout the Cenozoic, the contest between land and water continued to bring drama and flux to the European scene, as it had done throughout geologic time. Europe was still being gradually built up as a continent all through the Tertiary; and as she rose out of the seas, she was flooded time and again, as the waters rose and sank in cycles spanning many million years.

The transgression, or inundation, was at a peak in the Cretaceous, when the Chalk Sea covered most of the British Isles, France, Denmark, northern Germany, and southern Russia to the Urals. At the end of the Cretaceous there came a general regression and only remnants of the sea were left in the Paleocene. To the north the Scottish and Fennoscandian shields were high and dry, separated by the North Sea basin. The southern rims of the North Sea were still covered by water in a system of local basins: the Hampshire, London, Flandrian, Paris, and Danish-German basins. A seaway appears to have extended eastward, connecting the North Sea with the Volga basin.

South of the flooded areas, a Central European shield ranged from Brittany in the west to Podolia in the east. It rose to form highlands in the Armorican (Brittany) and Central French massifs, the Vosges, Black Forest, Rhine massif, Harz, and Bohemian mountains. It was bounded to the north by swampy lowlands and to the south by the great geosynclinal sea of Tethys, the then Mediterranean. A geosyncline is a technical term denoting a large elongated trough, which collected great masses of sediments over a long period; the strata thus formed may later become folded into mountains. The area which was to become the Alps was still deep under water, except for a few strings of islands beginning to push up out of the great expanse of Tethys.

In the southwest other seas encroached upon what is now southern France and the Iberian Peninsula. A deep gulf from the Atlantic lapped the foothills of the Central massif, and was connected to the south with the Ebro basin across what was later to become the Pyrenees. The Iberian peninsula, then an island, was also cut off from Africa by a trench from Tethys to the Atlantic, the Betic strait. With its old highland, the Spanish Meseta, the big island lay like a mountain fortress in the Atlantic.

This is Europe at the beginning of the Age of Mammals: a half-drowned continent, lapped by a glittering tropical sea. The old Caledonian and Hercynian ranges form the mountainous cores, surrounded by flooded lowlands and innumerable islands. The immense seaway of the Tethys opened into the warm waters of the Indian Ocean and carried them to the European shores. Monsoons, warm and laden with moisture, brought a humid tropical climate to much of the land. On the other hand, the North Sea region and the adjoining basins were connected with the cool waters of the Arctic (or Boreal) Ocean, and so counteracted the warming influence, at least until the time when, in the early Eocene, a Channel connexion to the west brought a stream of warm Atlantic water and accompanying southern species into the Anglo-Parisian Cuvette.

The Paleocene of Europe is divided into four ages. The earliest, or Danian, is represented by marine deposits mainly in Denmark and adjoining areas. We find here the remains of marine crocodiles, but the typical seagoing reptiles of the Cretaceous are gone. Next comes a sequence of marine faunas from the bay that covered the Paris basin and the Low Countries; this age takes its name, the Montian, from the Calcaire de Mons in Belgium. Most of the middle Paleocene or Thanetian is also marine, but land mammals of about this age are known from two areas: Walbeck in Saxonia and Cernay near Reims. Mammals of late Paleocene or Sparnacian date are known from the Paris basin, Belgium, and southeast England. These are our only glimpses of the Paleocene land animals in Europe.

The Walbeck fauna comes from a fissure filling in limestones northeast of Helmstedt in Saxonia. This assemblage is peculiarly unbalanced in a way typical of fissure faunas. It seems that the fissure served as a den for small carnivores and insectivores; remains of the former make up more than one-half of the total. Some primates and condylarths are also found, representing occasional occupants or prey brought in by the carnivores.

In addition to mammals, crocodiles and salamanders are found at Walbeck; the latter suggest that the climate in this high-lying part of the Central European shield was comparatively cool and continental, with numerous lakes. But the presence of crocodiles indicates that the climate was no cooler than what we would call subtropical.

Most of the material from Walbeck consists of loose teeth, of which there are many thousands. The majority belong to three species of primitive arctocyonids, which were technically condylarths but probably largely carnivorous in habits; they range in size from a fox to a jackal. The insectivores include both shrew-like and hedgehog-like animals, as well as some of the aberrant and probably amphibious pantolestids. All of these animals are closely related to North American forms and the same holds for the Primates, which belong to the Plesiadapidae and Carpolestidae, a group now often referred to the Insectivora. (It may be remembered that the latter were 'imitating' multituberculates, a group not found at Walbeck.) Finally, all the 'true' or herbivorous Walbeck condylarths belong to a family (the Hyopsodontidae) which had its heyday in the Torrejonian of North America: these were tiny, perhaps tree-living forms with clawed feet, not looking much like ungulates. The comparison between the Walbeck and the North American faunas indicates a late Torrejonian or perhaps early Tiffanian date.

A more detailed picture of the Paleocene fauna in Europe is given by the assemblage at Cernay. It is somewhat later in date, probably late Tiffanian. The fossil mammals are found at several localities, evidently representing the delta of a great river flowing from the east. At Mt Berru the freshwater bowfin (*Amia*) is a very common fossil; here we are still within the banks of a river flowing serenely westward. Further to the west, at Cernay, we get a whiff of the sea: sharks and rays prowled these reaches of the delta and left their teeth by the thousand in the deposits, together with the bones of carcasses brought out by the current.

The great river flowed through a low-lying land, much of which seems to have been covered with bush vegetation, interspersed with spinneys of higher trees. The rivers, ponds and lakes were bordered with gallery woods formed by willow, poplar, alder and elm, together with sassafras, ivy and laurel. Palms, alone or in small stands, would be seen from place to place. The climate was tropical, but with moderate rainfall. In the dry season the small ponds evaporated, leaving glittering gypsum deposits. Perhaps the coolness of the sea and the sun-drenched heat of the land conspired to keep the rains away from that coast.

In the warm, sluggish waters of the river, big champsosaurs up to four or five yards long, allied to the American forms, were common. Always there would lurk the danger of those long, tooth-studded, snapping jaws so oddly mounted on a broad, flat skull, with closely-spaced eyes. In comparison the crocodilians, though numerous, would seem nearly harmless; for, although occasional giants grew to eight feet, most were much smaller. Turtles of various kinds are found, the most common being a medium-sized species about a foot long. There are also salamanders and frogs.

Lizards were common on dry land, but no snakes at all have been found. Land tortoises about two feet long were also present.

The insectivores are reminiscent of Walbeck and many seem to be lineal descendants of Walbeck species. The same holds for the bear-like condylarth *Arctocyon primaevus,* which was the first Paleocene mammal ever to be described (by De Blainville in 1841). We now know the entire skeleton of this creature, which stood about eighteen inches high at the shoulders and was the largest mammal at Cernay. Its short, extremely sturdy limbs are of the basic condylarth type. The powerful canine teeth may have been useful in handling prey, but the skeleton indicates little agility or speed, and the brain was small and primitive. To a reasonably quick and alert animal, clumsy, dull-witted *Arctocyon* was no danger, and most probably it lived in bear fashion on fruits, berries and succulent plants as well as on meat. The great number of remains may indicate that *Arctocyon* was gregarious.

The only definitely carnivorous mammal at Cernay seems to have been *Dissacus,* a flesh-eating condylarth; but it may have haunted the shores in search of tortoises or shellfish rather than warm-blooded prey.

One of the most abundant mammals was the small condylarth *Pleuraspidotherium,* an animal some two feet in head and body length. It belongs to a family (the Meniscotheriidae) with advanced, crescentic cheek teeth almost looking like those of a miniature rhinoceros. With such a tooth pattern they probably were the most exclusive herbivores of their time; they may have formed large herds, browsing on the foliage of the riverside vegetation. In contrast the hyopsodonts so common at Walbeck were comparatively rare at Cernay.

Although these deposits are sometimes called the *Pleuraspidotherium* beds, the most common animal is a species of the primate genus *Plesiadapis.* This rodent-like form, about the size of a cat, moved in great hordes along the shores of the river, seeking food among the berries, nuts, and fruits.

So far the scene on land at Berru, away from the big reptiles of the river, would have seemed a peaceful one. But we have not yet met the strange and terrifying bipeds that moved about almost like the dinosaurs of the distant past. These were the diatrymas or terror cranes, *Gastornis* and other forms: great flightless birds, some of them reaching a height of seven or eight feet. They had reduced wings like the ostrich, and strong running legs; but there the resemblance ends, for the diatrymas had enormous heads with deep, powerful beaks, and the four-toed feet carried great claws. Towering over the mammals of its day, this giant bird was fast

Figure 7. The great terror crane *Diatryma* attacking a small condylarth, *Meniscotherium*. Lords of creation in the Paleocene, these rapacious ground birds later declined in competition with advanced carnivorous mammals.

enough to outrun them all and powerful enough to kill any one of them. It is almost as if the ancient reptile line was making a final bid for the supremacy on land; for birds, of course, are descended from reptiles, and are cousins to the dinosaurs and crocodiles.

The diatrymas probably arose from early wading birds, taking the opportunity to fill the niche for a great bipedal carnivorous being that had been vacated by the passing of the Mesozoic flesh-eating dinosaurs. In the later Paleocene we find them both in Europe and North America. They appear to have been the lords of creation at Cernay and were to remain a menace to the mammals well into Eocene times. Perhaps no other being that lived during the Age of Mammals, suddenly recreated in the flesh, would have made quite as nightmarish an impression as these imperious effigies of the roc bird.

But Cernay marks but a passing moment in geological time. While the river of Berru went on building its delta westward into the wide, shallow waters of the Paris basin, the faunas continued to evolve and change. As we move into the end stages of the Paleocene, the world of living things that inhabited the entire sweep of lowlands surrounding the Anglo-Parisian Cuvette seems gradually to come into focus. In Flanders, in the Paris basin, in southern England the bones and teeth of mammals begin to collect in the deposits. Walbeck and Cernay were two brief preludes; now, as we are leaving Paleocene times, the images of past life begin to crowd in. We are still in the Paleocene or Sparnacian, by European reckoning; but in North America the deposition of the Wasatch was already under way; so this story is better told in the Eocene chapter. And from now on, the story in Europe, as in North America, is an uninterrupted one.

Many students of Paleocene life have been struck by the similarity between some of the species found in Europe and North America; this must indicate that routes for continuous intermigration were available. Two main possibilities have been suggested. One is a direct route across the Atlantic, probably extending from the British Isles across Iceland and Greenland. The other is a route across the Bering Strait between Alaska and eastern Siberia; in that case, the animals would have to cross the whole of Asia.

If the latter alternative is correct, we must expect to find the common European-North American forms in Asia too. Unfortunately our knowledge of the Paleocene mammals of Asia is very poor, but there is one locality that has yielded Paleocene mammals. This is the Gashato Formation in the Gobi Desert in Outer Mongolia, a series of sands and clays with intercalated lavas, resting on dinosaur-bearing Cretaceous beds. The Gashato mammalian fauna apparently dates from the very end of the epoch, the Clarkforkian. It is a highly varied assemblage, obviously the result of a long, hitherto unknown history in Asia.

Looking at Gashato, one is struck by the almost systematic absence of the common Europeo-North American forms. We may note that the Gashato multituberculates (two species) are of the taeniolabidid type, which is absent in Europe; whereas the small ptilodonts common to Europe and North America have not been found at Gashato. There are no primates at all, and the sole true condylarth is aberrant, of unknown connexions; the same is true of the two insectivores. The carnivores include an arctocyonid and two creodonts of unknown affinities. All this is rather unsatisfactory and serves to emphasise how much we still do not know about the early history of the mammals in Asia. But as to the faunal exchange between Europe and North America, the facts are very plainly to be seen: it could not have taken place across the Bering bridge and so a North Atlantic route appears the most probable alternative.

Though European elements are lacking at Gashato, there are several forms with American affinities. In addition to the taeniolabidid multituberculate already mentioned, there is for instance an early member of the Dinocerata which is closely related to the Clarkfork dinocerates of North America. Even more interesting, and also puzzling, is the appearance of a member of the order Notoungulata for, except for an Eocene form in North America, these mammals are otherwise restricted to South America.

The presence of American forms suggests that a land bridge was in fact available across the Bering Strait, intermittently at least. But it is equally clear that there was no connexion at all with Europe, and we are reminded of the geological evidence indicating that an interior sea extended from the Arctic to the Tethys east of the Ural mountains.

Still another mammal found at Gashato, named *Eurymylus*, has long been thought to be a member of the order Lagomorpha – the hares, rabbits and pikas of the present day. The remains are fragmentary, but the presence of a double set of chisel-like incisors in the upper jaw would suggest a lagomorph relationship (the Rodentia only have a single pair). This might then mean that the lagomorphs arose in Asia. But the lagomorph affinities of *Eurymylus* have been doubted and there may be reason for regarding it as an aberrant member of the Insectivora instead.

Elsewhere in the world, Paleocene mammal life is known only from South America; and the history of this continent is a chapter of its own. In Africa, Paleocene deposits have so far failed to yield any fossil mammals, although fish, crocodiles and turtles have been found in phosphatic beds of this date, both in western North Africa and in the Niger basin. For many reasons, it would be of the greatest importance to know what the isolated Noah's Arks of Africa and India were carrying in the way of early mammals.

Let us now look back at the Paleocene of the World Continent as set down in Table 2. It lasted about ten million years, during which the mammals radiated and diverged from uniform, insignificant beginnings to form a varied and balanced fauna. While other groups of vertebrates show little if any significant change (except the birds, but their history is incompletely known), the mammalian fauna underwent a tremendous expansion. This is clearly shown by the number of distinct families known to have existed during successive ages, as set down in Table 3.

In the late Cretaceous we have so far identified thirteen families of mammals, of which six belong to the Multituberculata, one to the marsupials,

Table 2. The Paleocene of the World Continent

Epoch	North America (ages)	Europe (faunas)	Asia (faunas)	Date (Million years)
Eocene	Early Wasatchian	Sparnacian faunas		
				55
Paleocene	Clarkforkian		Gashato	
		Cernay		57
	Tiffanian			
		Walbeck		59
	Torrejonian			
				62
	Dragonian & Puercan			
				65
Late Cretaceous	(Lance)	(Maestricht)	(Djadochta)	

Table 3. Number of mammalian families in the late Cretaceous and Paleocene of the World Continent

	Late Cretaceous	Paleocene		
		Early	Middle	Late
Total number of families	13	17	27	41
First appearances	13	6	10	14
Last appearances	2	0	0	9
Percentage newcomers	100	35	37	34
Percentage extinctions	15	0	0	22

three to the Insectivora, one to the Primates, one to the Condylarthra, and one to the Creodonta. Two of the families (or fifteen per cent) became extinct at the end of the Cretaceous. With the addition of six new families in the early Paleocene, the total number increased to seventeen. There was no extinction on the family level. Ten more families were added in the middle Paleocene, bringing the total to twenty-seven families; again, none of them became extinct. Still another batch of fourteen families was added in the late Paleocene, so that the total now in existence was forty-one. So new forms were being added throughout the Paleocene, while there was very little extinction and the net result was a tremendous increase in the variety of the fauna.

Only in the late Paleocene does extinction increase, as nine families vanish from the record. All of these were archaic forms, belonging to the Multituberculata, Insectivora, Pantodonta, and Condylarthra; they were primitive mammals that were overtaken and crowded out by more progressive competitors.

Looking back in geological perspective, those ten million years of Paleocene time seem crowded with events and images. Evolution was going on at an unprecedented rate. Yet, of course, those ten million years in reality meant aeons of sameness; millions and billions of individual beings living out their lives under constellations that change too slowly to be perceived by earthly beings. It was an ever present now, of life and breath and browsing and hunting; only the actors were changing and dying; and a few left their bones where we find them, sixty million years later, as if carried to us on the back of time's arrow.

Chapter 3

The Eocene: epoch of consolidation

> The coast is straight and sombre, and faces
> a misty ocean. Red trails are seen like
> cataracts of rust streaming under the
> dark-green foliage of bushes and creepers
> clothing the low cliffs. Swampy plains open out
> at the mouth of rivers, with a view of
> jagged blue peaks beyond the vast forests.
>
> Joseph Conrad: *Lord Jim.*

THE GEOGRAPHY of North America had changed but slightly as the year-millions rolled by and Eocene time began. The Atlantic still encroached upon the east coast: Long Island and much of the coast south of it were submerged and the coastline formed a curve passing approximately through the present-day cities of New York, Philadelphia, Baltimore, Washington, Richmond and Norfolk. To the south, Florida and most of the present-day Gulf Coast were also flooded to about the same extent as in the Paleocene. So was the great peninsula of Yucatán, while Central America was gradually emerging as a string of islands. The deep Mississippi bay still encroached inland to present-day Cairo.

Like the Rocky Mountains, the Appalachians were worn down to an even surface before their rejuvenation; but this so-called peneplane stage was reached somewhat earlier, so that the area was a nearly even plain in early Tertiary times. Only a few peaks rose above it, for instance the Great Smokies and the White Mountains. Remnants of that old peneplane can still be seen in the flat surfaces of some of the highest ridges of the Appalachians; standing on them, we stand upon the surface of the Eocene plain and yet in a completely different landscape. After the Eocene, the region was gradually upwarped and as river valleys cut into the ground, the ancient mountains were reborn.

Looking now to the west, most of the Great Basin area was high ground in Eocene times and so this was a region of erosion, not deposition. It was not until the Miocene that basins developed here to receive sediment and preserve a fossil record of the life of those days.

Still further west the Sierra Nevada and the Coast Range were being built-up in paroxysms of earthquakes as folding and faulting proceeded. Between the two ranges, the California trough was invaded by sea water. The Sierra Nevada and the trough were created by the gradual tilting of an immense block, the eastern margin of which has been uplifted some 13,000 feet and now forms a forbidding scarp about two miles high; while the western part was slowly pushed down to some 25,000 feet below sea-level, so forming a trough.

The main mammal-bearing deposits of the Eocene were still being formed in the Rocky Mountains, where the intermontane basins were being gradually filled as the young mountain ranges were whittled away by the forces of erosion. Deposition within each basin was going on independently of the others, governed by local conditions. The record within any single basin is broken and incomplete, but the sequence can be pieced together from basin to basin to recreate a nearly complete calendar of Eocene times, spanning something between 10,000 and 20,000 feet in thickness of strata and some eighteen or nineteen million years of geological time.

Direct measurement of the time involved may be possible in at least one instance. This is the Eocene of the Green River Basin in Colorado, Utah and Wyoming. In Eocene times this basin formed the bottom of a lake called Fossil Lake, which finally covered more than 5,000 square miles; here were deposited some very fine shales with a high content of organic matter – the deposit even yields oil when distilled. Most of the folding in this area had occurred as late as the Paleocene and in the process a barrier was raised to the earlier, eastward drainage of the Colorado Plateau. The rivers now began to drain into the basin, the floor of which gradually subsided in the west. In this way Fossil Lake was formed, gradually becoming larger until by the end of the early Eocene it was a vast sheet of water covering large areas of Utah, Colorado and Wyoming, to a depth of probably more than one hundred metres.

The Green River shales have a delicate lamination of paper thin varves, formed by alternating calcareous (light-coloured) and organic (dark) mud, evidently representing a succession of dry and wet seasons. Remains of plants, insects and fishes are found here in almost incredible profusion; but of mammals there is very little, apart from a few footprints and – as an illustration of the millionth chance coming off – the wondrously preserved, fragile skeleton of the oldest known bat in the world.

If the varves represent the annual alternation of seasons, as appears probable, it can be calculated that the Green River Formation is equivalent to about 6,500,000 years. This would include about $2\frac{1}{2}$ million years of late Wasatchian (early Eocene) and four million years of Bridgerian (middle Eocene) time. On this basis it has been computed that the whole of the Eocene represents some nineteen or twenty million years, a figure confirmed by later results obtained by radiometric dating. This is a very good example of two quite independent dating methods leading to the same result and thus confirming each other. It also shows that the Eocene was the longest of the Tertiary epochs by a wide margin.

At the end of Bridger time, Fossil Lake dried out, probably because it became filled with sediment. Later sedimentation in this area was formed by traversing rivers gradually building up their beds.

The famous flora of the Green River deposits contains more than 135 identified species. Most of these are lowland plants that probably grew less than 1,000 metres above sea-level (the present altitude is more than twice as high). The flora includes ferns, cycads, an aquatic plant (*Potamogeton*) and palms, as well as willow, walnut, hickory, linden, alder, birch, hornbeam, oak, sumac, soapberry, golden-rain tree, tulip tree, sweet gum, maple, and many others. The higher flanks of the mountains were covered by pine forest. The flora grew under subtropical or warm-temperate conditions with a distinct alternation between dry and rainy seasons.

Another Eocene flora, probably approximately contemporaneous with

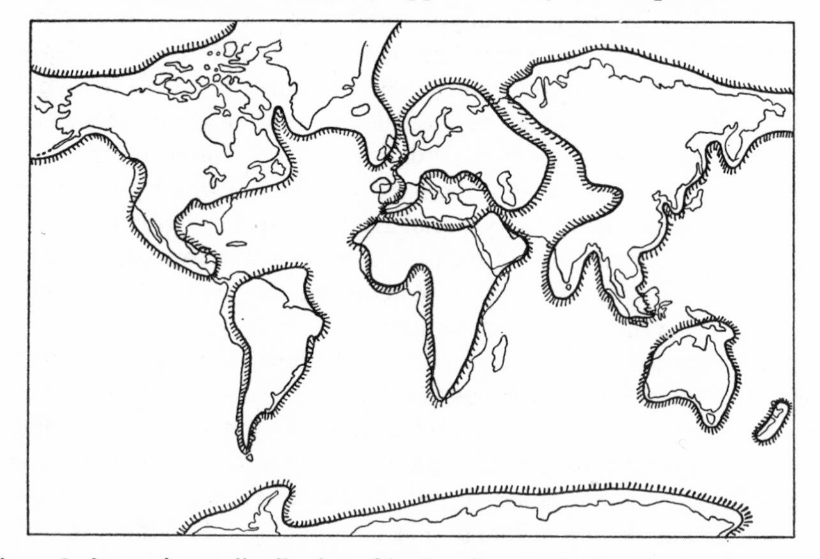

Figure 8. Approximate distribution of land and sea in the Eocene epoch.

that of Fossil Lake, comes from Chalk Bluffs in north central California and has a stronger tropical element with bread-fruit tree and the like. A relative of the modern Katsura tree (*Cercidophyllum*), a forest tree of southwest China, is common in this and other early Tertiary floras in North America. There is also an admixture of upland elements in the Chalk Bluffs forest, such as hickory, maple, and ash.

One of the most spectacular fossil floras that we know of is that of Yellowstone Park; its exact date is not certain but suggestions range from the Eocene to the Miocene. Here, petrified logs and stumps are found in layer upon layer, alternating with leaf beds. Big Horn Mountain in the northwest corner of the park is in part built almost exclusively of such fossil forests on top of each other in spectral succession.

In Eocene times this was a lowland, clothed with magnolia, walnut, platan and chestnut trees together with subtropical forms such as the fig and several species of the laurel family. But there were also pines here, and over all the others towered the giant redwoods, with trunks up to fourteen feet across.

Then the forest died in the swift agony of choking ash falls as the nearby volcanoes erupted; charred logs, falling every way, were buried in the ash. But as the centuries passed, the ash layer became covered with new vegetation, slowly building up to a climax until finally a splendid new forest lived over the grave of its predecessor.

Again the volcanoes awoke in response to their hidden pulse and once more engulfed the land in ashes; so the new forest died and was buried like the old. The cycle was repeated again and again, through the millennia, and the dead forests were heaped on top of each other so that we can count up to twenty-seven in succession.

Later on, siliceous water impregnated the wood where it lay buried, causing it to become harder than the surrounding ash. Finally, upwarping brought the whole area under the attack of erosion; and as the ashy matrix is swept away, the dead forests of forty million years ago emerge in an utterly different landscape.

Other Eocene floras in North America fill out the picture and speak consistently of tropical associations. The upper Eocene Comstock flora of Oregon is dominated by *Cinnamomum* (cinnamon, camphor tree) and a magnolia of a type now growing in the southern United States. Other Comstock species have their nearest living relatives in the Philippines, Mexico, Central America, and tropical South America. Again, the Wilcox flora at Puryear, Tennessee, shows tropical affinity and has a high percentage of entire-leaved dicotyledons like the modern rain forest of Panamá.

At the same time, there is a wide variation between these floras, reflecting the multitude of different environments over the continent – the mountains, the uplands, the coastlands, the swamps.

The lower Eocene is called the Wasatchian, from the Wasatch deposits studied by the pioneer field worker F. V. Hayden in southwestern Wyoming and neighbouring parts of Utah. Later on, more richly fossiliferous deposits of the same age were unearthed in other areas, for instance the Big Horn Basin and Wind River Basin in Wyoming, the San Juan Basin in New Mexico, and many other areas within the Rocky Mountains as well as the Colorado Plateau. The Wasatchian is now known to represent an immense length of time, some six or seven million years at least, and is often divided into two substages: the Gray Bull (lower Wasatchian) and Wind River (upper Wasatchian).

Middle Eocene or Bridger faunas come from the Green River and Bridger Formations of Wyoming, Colorado and Utah, and other sites. The Bridger Basin contains a remarkable series of successive faunas of this age. The sequence begins with lake deposits, followed by sediments laid down intermittently on flood surfaces, and is topped by ashes and tuffs representing a phase of volcanism on a grand scale. Excavation in these beds has yielded great numbers of excellently preserved fossil mammals, many of them in the form of complete skeletons. They are evidently of animals that were killed by the heat or gases from the eruptions. The Bridger stage is reckoned to have lasted about four million years.

The upper Eocene, in which the volcanism abated, is divided into two stages. They are the Uintan and Duchesnean, and both names refer to the Uinta Basin in Duchesne County, Utah. The Uintan probably lasted five million years and has left a fossil record in western Colorado, the Washakie and Wind River Basins of Wyoming, and to the north in Montana and Saskatchewan, in addition to the type area. In contrast the Duchesnean is little known outside its type area in the Uinta Basin and seems to have been somewhat shorter in duration – perhaps three million years.

Some Eocene localities outside the Rocky Mountains also contain good faunas, for instance the Sespe of California, but nothing to rival the great fossil fields in the intermontane basins. Marine deposits just off the Eocene coastline of the Gulf and in New Jersey and southward to Virginia give fascinating glimpses of the life in the seas, with some of the earliest whales.

The richness and variety of the mammalian faunas that populated this Eocene world is staggering and to do justice to it would require several books of this length. Here we can only take a brief look at the more remarkable types – those mammals that would surely have caught our eye had we been on a safari in the Eocene world.

Some orders of mammals were now nearing the end of their existence and among them were our old friends the multituberculates, whose long story ends with the Eocene. Another, less ancient but still archaic group of mammals are the large, rodent-like tillodonts; like the multituberculates, they did not survive the end of the Eocene. A similar fate befell the taeniodonts, of Paleocene origin. They were by now highly specialised

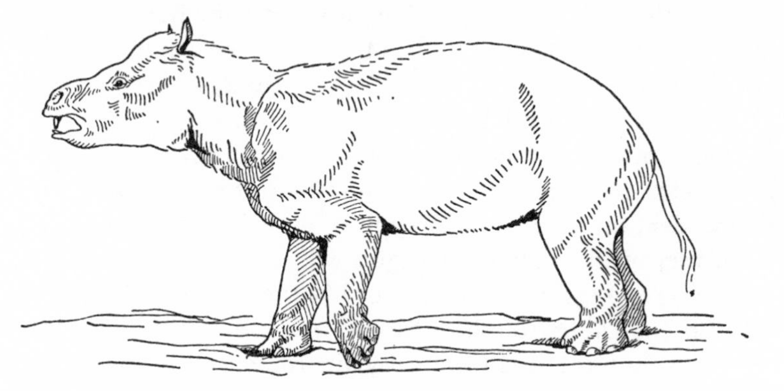

Figure 9. *Coryphodon*, a large hippo-like creature, measured about 2·5 metres from nose to base of tail. Appearing in the late Paleocene, it was especially common in the early Eocene of North America and Europe; it is in the order Pantodonta.

animals, somewhat resembling the giant ground sloths that were to evolve many million years later. The Bridger and Uinta *Stylinodon* represents the evolutionary climax of this aberrant line. In this animal, both the canines and cheek teeth had open roots and grew continuously, which necessitated the development of extremely deep, heavy jawbones. This is an early specialisation for heavily abrasive food, but it does not seem to have been particularly successful, for taeniodonts were never numerous and the last died out at the end of the Uintan.

The Pantodonta, which had long been the giants of their world, reached their acme in early Wasatchian times with the great *Coryphodon*, common both in North America and Europe. It was a short-legged, heavy beast about three metres long, not counting the short tail. Its big head was carried on a rather long neck; the eye-teeth were tusk-like. The coryphodonts probably resembled hippopotami in habits as well as appearance, for their weakly built spine suggests that the long body was normally submerged in water, which would buoy it up without the need for strong muscles and ligaments. The coryphodonts had very small brains, although the peculiarly raised crown of the head may have given them a spuriously intellectual appearance.

The pantodonts were superseded as upland giants by the uintatheres or Dinocerata, archaic beasts of bizarre and ungainly appearance. Their heyday was the Bridger and at the end of the Eocene they were all gone. The typical uintatheres carried three pairs of bony protuberances on their heads: a front pair on the tip of the nose, a second on the forehead, and a third pair at the back of the head. The two hind pairs were joined by fore-and-aft bone ridges enclosing a sort of basin on the top of the skull. The upper canines were enlarged, sabre-like structures, protected by a flange on the lower jaw, and were present in both sexes.

Like the pantodonts, uintatheres had very small brains, but their feeding mechanism was somewhat more specialised. Even so, they must have been dependent on an abundant supply of succulent herbage.

The history of the uintatheres begins with small, hornless forms in the late Paleocene and early Eocene. Forms like the Bridger *Uintatherium* and the late Eocene *Eobasileus* show progressive size increase; the last-mentioned was a giant as big as the great white rhinoceros of the present day. By Duchesnean times the uintatheres were gone, perhaps because of a gradual drying of the climate which would affect the vegetation on which they browsed. Or perhaps their feeble brains made them unfit for survival in a world when efficient true carnivores were beginning to appear.

The Condylarthra was another archaic order which tended to decline during the Eocene. The semi-carnivorous arctocyonids straggled into the Wasatchian, then died out; the mesonychids – a second family of carnivorous condylarths – held out better and survived into the Oligocene. The central type of condylarth is well represented by *Phenacodus*, still a common form in the Eocene. With its long skull, long and flexible body, and short, sturdy limbs, it was a typical early ungulate, probably rather close to the origin of the modern hoofed animals. Its fore and hind feet still have the five original digits, terminating in small hoofs. This animal was adapted to moving in the forests and on soft ground, while its performance on the plains would have been much inferior to that of a modern gazelle or donkey.

A very small and imperfectly known form in the Wasatchian seems to be a member of the Notoungulata, a South American group. We have already met another representative of the notoungulates at Gashato, in the late Paleocene of Mongolia. It has been thought that some member of this order succeeded in migrating northward from South America across the

Figure 10. Perhaps one of the most grotesque animals that ever lived, the dinocerate *Uintatherium* flourished in the middle Eocene of North America. Horns and sabre-like canine teeth were the chief means of defence of this rhinoceros-sized animal.

Figure 11. A central type in the Condylarthra was *Phenacodus*, which ranges in time from late Paleocene to middle Eocene, and lived both in North America and Europe. With a head-and-body length of 5·5 feet (about 1·7 metres) it was one of the larger condylarths.

straits of Central America, probably being rafted from island to island – a common method of involuntary travel for tiny mammals. In this way the notoungulates got a chance to colonise parts of the World Continent, but their success was indifferent and they soon became extinct outside South America.

So far we have reviewed the old-timers in the Eocene, forms now long extinct. Let us turn now to the newcomers, which were arising in profusion to dispute the reign of the continents with the archaic mammals of the Cretaceous and Paleocene radiation. The late Cretaceous and Paleocene had been a time of great innovation and productivity. Some fifteen orders of mammals arose in that radiation within the World Continent. But they were not destined to last; more than one half are now extinct.

The Eocene production was more modest but also more consistently durable. All of the seven orders that make their earliest appearance in the Eocene are still with us today. The appearance of new orders abated after the end of the Eocene. Only a few orders make their first appearance in Oligocene strata, and most of them probably really arose in pre-Oligocene times. So the radiation of the modern mammalian world took place in the Eocene, and the basic diversity of the living mammalian fauna had been established at the close of that long, eventful epoch.

Both of the great orders of modern ungulates appear early in the Eocene and presently flower into an enormous variety of types, many of which were large and impressive and eclipsed the archaic giants. Let us first look at the Perissodactyla or odd-toed ungulates, to which belong the horses, rhinoceroses and tapirs of the present day. This order produced no less than thirteen families in the Eocene (a fourteenth appeared in the Oligocene). In later times, extinction reduced the variety; the later history of the Perissodactyla is one of consolidation of a few successful lines, and finally of retraction and gradual extinction. Modern perissodactyls are definitely on the wane.

Horses of a kind, or rather ancestors to the horses, are the earliest perissodactyls to appear in the fossil record; as we have seen, the earliest appeared before the end of the Paleocene. Probably no one seeing these fox-terrier-sized animals in the flesh would take them for horses, and when Richard Owen discovered and described them in 1840 the idea of such a relationship did not enter his head. He noted some resemblance to the dassies or hyraxes and named the animal *Hyracotherium*; it was only a good deal later that O.C. Marsh discovered the sequence of intermediate forms linking *Hyracotherium* with modern horses. He then gave it the name *Eohippus* or dawn horse. However, a scientific name once given cannot be replaced by another even if more 'appropriate', so we can only use eohippus as a vernacular name.

Little eohippus was much more like a condylarth than a horse. It had the long, flexible body, elongate head, and complete dentition typical of archaic ungulates. The legs were shorter in relation to the body than in a horse, but the hind legs especially were longer than usual in condylarths. The feet, unlike those of condylarths, were held up from the ground, so that the animal walked on its toes; this type of gait is called digitigrade. In addition to hoofs the toes also had pads on the under side as in a dog, and the weight was spread between the pads and hoofs. The number of digits was reduced to four on the fore-feet and three on the hind feet. The brain, though still lowly for a mammal, was definitely superior to the ordinary condylarth type.

Yet this agile little animal was utterly different from a modern horse,

Figure 12. Dawn horse or eohippus, *Hyracotherium*, from the late Paleocene and early Eocene was the first of the odd-toed ungulates. Most species of this genus were half a metre or less in length.

which has a head of quite another shape, a highly specialised battery of cheek teeth, a long toothless diastema, a straight and rigid backbone, and long legs each terminating in a single, powerful toe with a hoof but no pad.

Eohippus was present in Europe as well as North America, but the subsequent mainstream of horse evolution is localised in North America, though from time to time migrant forms reached out from that centre to colonise other parts of the World Continent.

From eohippus of the Wasatchian we can trace a lineage passing through the genera *Orohippus* of the Bridger and *Epihippus* of the late Eocene. The size remained static, varying in different species from about 25 centimetres at the shoulders to almost twice that height. The teeth also remained low-crowned, but their shape was gradually altered, so that the premolars became more and more like the molars, a process called molarisation; and the cusp pattern on the cheek teeth developed into a number of crests or 'lophs'. But all the Eocene horses were evidently forest-living, leaf-eating forms, and no one seeing them could reasonably have foreseen that they were to turn into the big, fast, plains-living grass-eaters that represent the family today.

Related to the horses were the brontotheres or titanotheres, which evolved in a quite different way, to become gigantic, grotesque animals, gradually replacing the uintatheres in this arena of life. Where the rhinoceros carries its nose 'horn', they sported a diverging pair of bony protuberances. Their culmination came in the Oligocene; the Eocene forms were smaller and, at first, hornless.

There are several families of animals looking like primitive tapirs or rhinoceroses. Some of the latter were lightly built and obviously fleet of foot and are sometimes called 'running rhinos'; in the flesh they may have resembled early horses. Another group of rhinoceroses that appeared in the late Eocene, the amynodonts, were amphibious, hippopotamus-like animals. In that niche they obviously succeeded the coryphodonts previously described. Finally, true rhinoceroses of the modern family appeared as early as the Bridger, but they did not become numerous until the Oligocene.

The even-toed ungulates or Artiodactyla were even more varied than the perissodactyls in the Eocene, for they came to number no less than eighteen families in that epoch (the total number of artiodactyl families known is about twenty-nine, and the variety of this order of mammals is exceeded only by the Rodentia). Although the artiodactyls never showed such a spectacular array of large forms as the perissodactyls, they have in the long run been more successful than any other ungulates, and now make up the great majority of living hoofed animals. The secret of the artiodactyl success seems to lie in their extremely efficient limb construction, especially the tarsal joint of the hind leg. The artiodactyl astragalus bone – the one that joins the heel and middle foot to the lower end of the shin – forms an ankle joint of unusual mechanical efficiency. This type of bone is found fully developed in the earliest artiodactyls and can be shown to have evolved from a type seen in certain Paleocene condylarths.

Like other orders the Artiodactyla originated as small, unspecialised forms, which range in time from the early Eocene to the Oligocene. Some of these still had five-toed feet. They show the gradual evolution of the crescent-shaped (or selenodont) cusps so characteristic of the artiodactyl cheek teeth.

There soon appeared a variety of more or less pig-like forms, whose teeth suggest a mixed diet like that of the living swine. Other artiodactyls foreshadow the Ruminantia, which have a many-roomed stomach and chew the cud. In one group, which retained five digits on the fore feet, claws were present instead of hoofs. Some of these animals, which are called agriochoeres, became as large as a cow; they may have used their claws to dig for tubers. Other artiodactyls seem clearly enough to be primitive camels.

The squirrel-like *Paramys* and other primitive rodents were common in the Eocene, especially as the multituberculates became rarer in the middle and late Eocene. Several new types evolved towards the end of the epoch, among them the pocket gophers. Hare-like animals, on the other hand, are decidedly rare; this order (Lagomorpha) makes its first appearance in the late Eocene.

Turning now to the carnivores that preyed on these plant-eaters, we may note that the creodonts now dominate the scene. Prominent among these primitive carnivores were the families Hyaenodontidae and Oxyaenidae, both comprising many kinds of powerful, rapacious beasts. Like the true Carnivora, the creodonts tended to develop enlarged carnassial cheek teeth, but they were situated further back in the tooth row.

The true early carnivores, or Miacidae, were becoming specialised on divergent lines, one of which grew to the size of a small hyena and may have had the same habits. The other miacids remained small and foreshadowed the modern types of carnivores. In the late Eocene, representatives of several modern carnivore families occur, either in Eurasia or North America: the Felidae or cats; the Canidae or dogs; the Mustelidae or weasels; and the exclusively Old World family Viverridae, the civets and allied forms.

From some unknown insectivorous ancestor arose the bats, which suddenly appear in *Icaronycteris* of the early Eocene. They probably originated as gliding forms, but of the early stage in their evolution we have no record whatever, for the Eocene bats already have well-developed wings. In the middle Eocene, bats occur in both the Old World and the New. Several modern families of bats appeared in the middle and late Eocene, including, for instance, the horseshoe bats and the extremely varied vespertilionid family.

A great number of Primates are known from the Eocene; all of them belong to the 'prosimian' grade. We probably tend to think of this order in terms of apes and monkeys – clever, 'four-handed' beings with semi-human faces. But the order also contains a number of less man-like forms, mainly the lemurs and tarsiers. These prosimians or lower primates are nowadays restricted to limited habitats in the Old World tropics, but in the Eocene they ranged over the World Continent and were *the* Primates; the appearance of apes and monkeys was still in the future.

The Paleocene primates were somewhat rodent-like in appearance; many of them survived in the early Eocene, but they were then largely superseded by more lemur-like and tarsier-like forms. The important Eocene lemur family Adapidae was present both in Europe and North America. The European forms (with *Adapis*) appear to have been more specialised fruit-eaters, while the American *Notharctus* and its relatives represent a more generalised type, probably with a mixed diet. These forms may be fairly close to the ancestry of the higher primates.

The origin of the Eocene lemurs is obscure; the known Paleocene forms

Figure 13. The prosimian primate *Notharctus*, a lemur-like animal from the Eocene in North America; length without tail about eight inches (20 cm.). With its powerful hind legs and prehensile hands it presumably was an agile and fearless climber.

are too specialised to be ancestral. Possibly the adapids originated in Africa; we cannot be certain as long as the Paleocene mammal fauna of that continent is unknown.

Some other small prosimians from the Eocene appear to be related to the living spectral tarsier, and they even have the enlarged eyes and elongated hind limbs so characteristic of this tiny, nocturnal insect hunter, which is able to make standing jumps of amazing length. But many of the early Tertiary prosimians float about in a systematic vacuum and are the despair of classifiers. Their study is further hampered by the poor condition of much of the material, or by the irritating fact that the teeth may look much alike in quite distantly related forms. Some of the Eocene forms in North America are evidently ancestral to the living monkeys of South America.

The birds, like the mammals, had achieved their essential modern diversity by Eocene times, while some archaic forms now made their last appearance. The giant diatrymas or terror cranes flourished in the Wasatch, but their downfall was swift. Thus passed the last bid of the reptilian line for the rule of the World Continent. (Island Continent terror birds lasted longer.) Instead, we meet flying birds belonging to the modern orders – loons, petrels, cormorants, herons, ducks, vultures, cranes, gulls, owls, woodpeckers, and ancestral perching birds.

Then there are the peculiar saw-toothed birds, the Odontopterygiiformes, large birds somewhat resembling pelicans but with pointed 'teeth' in their jaws. On closer inspection they turn out not to be real teeth at all but spiky outgrowths of the jawbones themselves. These birds are rare and imperfectly known, but there are records that prove their survival into Miocene times.

Streams and lakes were still reptilian territory, ruled by champsosaurs and crocodilians like those of the Paleocene. Amphibious turtles like the soft-shelled turtle (*Trionyx*), marsh turtles and terrapins were also common and lived side by side with types now wholly extinct. The land lizards include iguanas, agamas, slow-worms, monitors and others, while the snakes are represented by forms related to the boas and pythons. Large sea-snakes are quite common.

The land bridge across the North Atlantic was still open in the early Eocene; let us cross it and take a look at the European scene.

The Eocene clearly was a time of warm climates. Coral reefs, which are limited to the tropical seas, are found in the Eocene of Europe. The seas swarmed with a remarkable one-celled animal belonging to the class Foraminifera. Most forams are microscopic or nearly so, but at some times in the history of the earth there flourished giant types, attaining sizes that seem fantastic for a unicellular being. One such group were the nummulites. They were so common in the earlier Tertiary that this part of the period is sometimes – especially by French students – called the Nummulitic. They have coin- or lens-shaped tests usually a centimetre or so in diameter, but occasionally up to two or three inches. A nummulite limestone with great concentrations of these tests may look like a carelessly heaped pile of small change.

The nummulites are not entirely extinct – one genus (*Heterostegina*) is still in existence and shows that the giant forams prefer shallow seas in the tropics and subtropics. In the Eocene, nummulites flourished in the European waters (they are also found in America, for instance in Florida); when the Anglo-Parisian Cuvette became connected with the Atlantic, it was invaded by nummulites in great profusion, showing that the reign of the Boreal Sea had come to an end. At the same time the flora of the London Clay underwent a change.

This flora carries us to a world of heat and humidity like that of the modern Malayan jungle. Nearly one-half (43 per cent) of the plant families represented are exclusively or mainly tropical in modern occurrence. Another even larger group (46 per cent) are evenly distributed in tropical and non-tropical areas today; and only a small number (11 per cent) are mainly extra-tropical at the present day. Many of the forms are now found in the coastal region of Malaya, and some are brackish-water plants. The inference is that southeastern England at the time was a low-lying, well-watered area traversed by many streams. The mean annual temperature has been estimated at about 21° C; today it is 10°.

Abundant rainfall may also be suggested by the production of lignites in Eocene times. The German Eocene brown-coal deposits are estimated at some 13,000,000,000 tons. But it should be noted that the lignite-producing woods grew in closed basins that could have been kept moist even by moderate rainfall. And in fact the climate must actually have been dry in some areas, for instance in the late Eocene in the Paris basin and in the northern forelands of the Pyrenees, where thick strata of gypsum were formed, and in the Ebro basin and the upper Rhine valley, where potassium salt was deposited. This indicates a gradual desiccation during the Eocene, enhanced in regions where the new, rising mountain chains created barriers for the rain-laden winds: the Vosges and the Pyrenees.

The Anglo-Parisian basin saw a succession of inundations and regressions, as the year-millions of Eocene time rolled by. Rising waters characterise the beginning of the Eocene, when the Ypresian clays were laid down. A regression at the end of the Ypresian left much of the lowlands around the basin dry, only to be covered again by shallow waters as the Lutetian transgression set in. But late in the Lutetian (or middle Eocene) the water began to withdraw once more, and the Paris basin now emerged from under the sea, to become a lake-dotted, gradually drying, low-lying plain. Towards the end of the Eocene definite eras of drought can be identified from the gypsum layers of Montmartre with their famous fossil mammals, studied by Cuvier more than one and a half centuries ago. It was here that he laid the foundations of the science of fossil mammals, and the area has remained one of the most prolific sources of Eocene fossils.

Much of our knowledge of Eocene life comes from the lignite-bearing beds in central Germany. Another highly productive type of fossiliferous deposit is the 'Bohnerz' or brown iron-ore carrying, clayey and sandy matrix found in many ancient caves and fissures. The limestones at Causse in southern France are riddled with such ancient fissures, now decorated with phosphatic stalactites and with a filling of reddish, phosphate-laden cave earth. These are the renowned Phosphorites of Quercy, which have yielded immense numbers of fossil bones. Unfortunately the early collectors omitted to make a stratigraphic separation of the material they excavated. It has later been discovered that the fauna is part Eocene and part Oligocene. Other bone-bearing cave earths of the Bohnerz type are found in the limestones of the western Alps and the Swiss and Swabian Jura.

The Alps were still being born, children of the Tethys. As soon as the first mountain arches reared their heads above the water, they were attacked by erosion. As a result there accumulated the widespread, marine Flysch deposits, consisting of darkish shales with limestone bands, sandstones and conglomerates in irregular succession. Little of the rich animal life that teemed in the cool, deep, oxygen-laden waters of the Flysch Sea has come down to us; there are just some occasional burrows made by nereid worms. The dramatic fluctuation between coarse gravels and fine-grained clays, without any discernible rhythm, is typical of the unstable sedimentation processes around the waxing mountain range. The Flysch Trench ranges all along the Alps in a great arch from west to east and continues along the Carpathians.

The basins on the Atlantic coast saw a succession of rises and falls of the sea. Meanwhile the Pyrenees began to emerge. In late Eocene times the southern Pyrenean trench was filled up and dry. On the other hand, to the south, the Ebro basin was subsiding slowly. The Sub-Betic and South Rif Straits still effectively isolated the Iberian island from Africa, but the land mass between the two straits may at times have been connected with Corsardinia, a big island incorporating parts of Corsica and Sardinia, and perhaps also in contact with the rising Pyrenees.

Further south a gulf of the Tethys, the Saharan Gulf, covered most of the Tunisian peninsula and extended far across the interior of the present-day desert. To the east, the northern parts of Egypt were still submerged in the earlier half of the Eocene, but the land was gradually being built up from the south. At Fayum we find the record of this advance of the land. The sequence begins in the late Eocene with a marine fauna, where sharks and early whales are common. By the latest Eocene the deposits become brackish or of freshwater type and contain masses of silicified tree trunks as well as a great mammalian fauna.

The European land fauna of the early Eocene, or Ypresian, is known mainly from the London basin and the related deposits of France and

Belgium. Its similarity to the North American Wasatch is striking, and it might even be said that Europe, in the early Eocene, was a province of the North American faunal region. We recognise such forms as the tillodont *Esthonyx,* the condylarth *Phenacodus* and the pantodont *Coryphodon,* all of which probably immigrated from North America. Little eohippus (*Hyracotherium*) is also found in both Europe and North America, but we do not know in which area it originated.

Of course there are also some forms that did not migrate. The brontotheres, which were just starting their long and spectacular career in North America, did not enter Europe until much later, while the North American semi-tapirs failed to do so at all. On the other hand, the palaeotheres and lophiodonts so characteristic of the European fauna did not reach the New World. The palaeotheres were a side branch of horse-like animals, while the lophiodonts were a European group of semi-tapirs. Both groups tended to develop larger forms as Eocene times passed by; the lophiodonts died out after the middle Eocene, but the palaeotheres culminated in giant forms as large as a great draught horse. Cuvier discovered these strange beasts in the late Eocene *gypse de Montmartre* in the eighteenth century.

Lutetian mammals in Europe are known from many sites, but the most remarkable of them all is the Geiseltal, some twelve miles south of the town of Halle in Germany. Probably no site in the whole world gives us such a picture of the living past. The Geisel itself is a little river, which originates as a spring in the township of Mücheln; it now flows sedately, but the time when it issued as a metre-high spout is still in living memory. The valley of the Geisel is the site of great brown-coal mines, the coal or lignite being the residue of peat beds that formed here in the swamps and bogs of the middle Eocene. The peat formed in large, closed basins, which were gradually subsiding. As the basin floor sank, the peat beds were built up and now form coal seams with thicknesses up to 100 metres and more.

Remains of plants and animals are preserved in the lignite with amazing detail. There are many sites in the world where we may find calcified or silicified tree trunks; but it is only in the Geiseltal that we also find the skeletons of the tree-living animals that were killed by the fall of the tree! Microscopic examination of the fossils may reveal the striated musculature of fishes and lizards; bacteria in the tracheae of beetles; the filigree network of fibre-shaped fungi; the cells and stomata of leaf cuticles. Flowers, leaves, dragon-flies are preserved; beetles may still glow with their original colours; muscles, intestines, and even the original contents of the beetle stomach (with symbiotic or parasitic fungi) may be studied as in a modern specimen freshly prepared.

We may find the shed skin of a lizard, perfect down to the 'glove' of horny scales that once encased a living foot. Miraculously preserved, there is the original cell structure, complete with cell nuclei, of the soft, moist skin of a frog. There are also the brains and spinal cords of frogs, preserved as calcium compounds of fatty acids. In a dead lizard, arteries traversing the petrified musculature are still filled as if with blood, and the blood-corpuscles are to be seen in this forty-eight-million-year-old preparation as clearly as in a modern one.

Here an india-rubber tree has shed droplets that became vulcanised by the sulphur in the deposit and preserved. There, the carcass of a small mammal has been petrified with part of its original skin and hair cover, and for once we may use 'external characters' in the classification of a fossil mammal. There again, in the stomach of a fish, we find the fossil of a worm-like being telling a strange story. It is a typical insect parasite, *Gordius*, and it must have been swallowed together with its host shortly before the death of the fish.

Add to this that the Geiseltal lignite in some cases has preserved, for forty-eight million years, the chlorophyll that gives a green plant its colour, and remember that the green colour is the first to fade from a dead leaf in the autumn!

In Lutetian times much of the land between the Vosges in the west and the Bohemian Massif in the east was probably a sun-scorched steppe. Only the river valleys and basins retained sufficient water to carry a rich vegetation. There was a strong seasonal variation between rains and drought; this is shown by the tree rings and also by rings in the scales and ear-stones of fishes. The brown-coal basin would stand out as a compact forest island in the steppe. On nearing the forest wall, the steppe would gradually come alive. Groups of palaeotheres might be seeking the shade in the palm groves forming the outer fringe of the wood. With them might be seen some pig-like animals called anthracotheres. A fast-running bustard was a true child of the steppe and only occasionally took wing to visit the waterholes in the basin. Solitary, lean hyaeonodonts might be sniffing on the track of an anthracothere herd, or awaiting the coolness of the night in the dense shade of the redwoods.

The mighty sequoia is the characteristic tree of the outer gallery forest surrounding the basin; but there are also pine, chestnut, oak, walnut, magnolia, fig and many others. There are great numbers of birds, but very few of their remains have come down to us. One species that we happen to know is a hornbill, a member of an exotic, weird-looking group now confined to the Old World tropics; other remains indicate the presence of a

Figure 14. The typical genus of the hyaenodont creodonts, *Hyaenodon*, appeared in the late Eocene of North America and Europe, entered Africa at the end of the epoch. Late forms of this wolf-sized animal are found in the Miocene.

small owl, various other small birds, and birds with large ornamental feathers.

Python-like snakes wind around the branches. The biggest grew to a length of 230 centimetres, but smaller ones have also been found, down to newly hatched specimens. Long-tailed agama lizards, brilliantly coloured, jump in the trees. Tree frogs are also present; as night falls they will join in the choir of their swamp-living or burrowing cousins.

There are mammals in the trees, too. Both lemurs and tarsiers are plentiful. A small opossum is nonetheless a giant compared with tiny *Ceciliolemur,* an insectivore less than ten centimetres in total length, of which more than half is taken up by the long tail. The hands and feet have long, clasping digits without well-developed claws; and remains of the pelage show that *Ceciliolemur* carried a sparse coating of spines in its fur.

As we move further into the basin, the wood gradually changes in character. Inside the forest wall we are in a region with palmetto brush, bog-myrtle, holly, willow and india-rubber trees. Here are also many larger and smaller ponds with muddy water. They are of karstic origin: the peat beds here rest upon sandstones that may in places become soaked out by percolating water. The collapse forms a funnel-like depression, which becomes filled by water from the overflow of the lake in the rainy season. In the dry season, the water evaporates or soaks away. The ponds may now form veritable death-traps for animals that are coming down to drink and, having tumbled down the steep banks, are unable to get out. In other ponds crocodiles are lurking. High above, where the entire system of lake, bog, and forest wall can be surveyed, and the flash of water is to be seen in every direction, a condor is serenely patrolling the air on broad black wings, ready to swoop down at the first sight of an animal in distress.

Here is the hunting-ground of the rapacious panzer crocodile, a long-snouted beast up to seven or eight feet in length. It was an odd-looking reptile with almost hoof-like claws and a whiplash tail quite different from the deep, compressed tail of normal swimming forms. Evidently this species spent most of its time on land, and indeed its heavy armour of bony scales might have been cumbersome in a swimmer. The panzer crocodile probably haunted the shores of the ponds and preyed on the animals that came down to drink. (See Plate 2.)

Some places seem to have been special 'eating sites', and we may picture a writhing swarm of crocodiles feasting on the big cadaver of a tapir-like lophiodont; here we find great amounts of crocodile excrement, teeth and bones. At other places the eggs of crocodiles occur, some of them with a partly developed embryo inside. A freakish find is a rolled-up female crocodile still carrying five eggs within her body. But sometimes all we find of a crocodile is just a bunch of pebbles that the animal had swallowed to help break up the food; they show the etching of the stomach acids, but of the animal itself nothing remains.

In the dry season the glades in the bog forest would be humming with insect life in the motionless, steaming air, with the dark wall of the gallery wood in the background. Dragon-flies sweep the air in regular beats; crickets, grasshoppers and cicadas form a many-voiced choir; a cloud of flies rises from the shore of a nearby pond as a crocodile moves up to a carcass and starts dragging it along. An entomologist hunting in the glade would also catch cockroaches, termites, neuropterans, homopterans, and an endless array of beetles: there are buprestid beetles, mealworms, glow-worms, chrysomelids, sexton beetles, ground beetles, and the beautiful water jewels (*Donacia*). On the other hand, no wasps or butterflies have so far been discovered in the Geiseltal, although there is little doubt that they did exist here.

The terrestrial animals moved between the gallery woods and the lower bush vegetation. The largest were tapir-like lophiodonts, animals about eight feet long and three feet high at the shoulders. They lacked the proboscis of the true tapirs, but otherwise their appearance (and probably their habits) would resemble that of modern tapirs. Fossil excrement or coprolites of lophiodonts have also been found; remains of plants in the droppings include some grasses, showing that the lophiodonts would graze on the steppe from time to time, but mostly they seem to have browsed on the leaves in the forest. They may have lived in herds or family groups. Sometimes apparently a lophiodont was killed by the drought; there are some specimens that evidently became mummified, lying in a dry water-hole, before they finally became covered by new sediments. Great numbers of bluebottle grubs are sometimes found in the lophiodon bones, especially the nasal cavity.

Palaeotheres also populated the Geiseltal in herds; they apparently had their selected water-holes and routes, for their remains are concentrated to a few 'horse cemeteries'. The Geiseltal palaeotheres are small but rather powerfully built, measuring some seventy or eighty centimetres (2½ feet) in length and forty centimetres at the shoulder. The palaeothere dentition is primitive like that of eohippus; the eye-teeth are enlarged and probably were used in defence. The feet, too, are primitive, with four digits in front and three behind, as in eohippus; with these spreading feet the palaeothere would move easily over boggy ground.

Several species of rodents lived in the Geiseltal, but they are incompletely known. Most of the carnivores were creodonts of the family Hyaenodontidae, but there are also a couple of miacids. By nightfall the big hyaenodonts may be imagined padding along the forest wall, while glow-worms and fireflies flicker in the dark and the mighty chorus of the frogs fills the air. Above, noiseless fluttering shapes may be vaguely made out from time to time: the Geiseltal bat, *Cecilionycteris*.

Many lizards related to the slow-worms and the iguanas slither through the underbrush. Some of the lizards are so heavily armoured as to look almost like miniature crocodiles; others resemble the slow-worms and among these there are nearly modern types without any legs at all, side by side with more primitive forms with short limbs. Land tortoises are also seen.

As we move towards the lake, a new, more open forest type appears, dominated by *Nyssa* (a relation of the pepperidge or sour gum tree) and the swamp cypress, *Taxodium*. Here the alligator is king and the small land animals keep out; only the big lophiodonts and larger carnivores brave the danger. There are many different kinds of alligator, most of which grow no larger than four feet, but occasional specimens reach the seven or eight feet that seems to be the maximum length for the early Tertiary crocodilians in Europe. All of them are broad-snouted, some indeed have a broader and shorter head than any living type of alligator. There is no true crocodile in the lake; the panzer crocodile is the only representative of this group at Geiseltal.

Swamp turtles of the genera *Ocadia* and *Geoemyda* are among the most common finds from this zone of the Geiseltal, but there are also a few remains of river turtles, suggesting some connexion between the lignite basins and the river systems of the day. But they are rare, especially in comparison with the situation at Messel somewhat further to the west, where there was a direct connexion with the rivers. Similarly, crayfish are rare in the Geiseltal; they too prefer less stagnant waters. But the lake and ponds probably were a fertile breeding ground for the malaria mosquito (*Anopheles*), of which a pair of maxillae has been found.

Finally the expanse of the central lake opens up, fringed by reed and lotus banks, with water-lilies and other fresh water vegetation. We know some of the fish that lived in the lake: three species have been found. One of them is related to the living salmon but had a shorter, deeper body with a large fin on the back. A primitive member of the pike family had much shorter and less bloodthirsty looking jaws than the modern form. The third species, like the other two a small, carnivorous fish, is a relative of the perch. High water would carry the fishes into the funnel-like karst holes, where they would dry out and die at the end of the rainy season; this is where their remains are now found, often in great numbers. Other hazards threatened those that remained in the lake; apart from the fish-eating alligators they were also preyed upon by the wading birds, among which a primitive crane is known to us.

How long was the Geiseltal basin active? In the lower part of the coal deposit there is an alternation between blackish and lighter-coloured bands, assumed to represent the dry and rainy seasons respectively. If a double cycle of this type corresponds to one year, the total time consumed in the deposition of the three lignite beds in the 'Cecilia' and 'Leonhardt' mines may be estimated at some 120,000 years. The deposition of the entire Geiseltal coal series would have taken at least twice as long, or about a quarter of a million years. This is a fair part of the early Lutetian, though only a small fraction of the age as a whole, for it probably lasted about four million years.

The Lutetian fauna in Europe already shows an incipient divergence from that of North America and the resemblance continued to fade. The connexion between Europe and North America had apparently been lost. The late Eocene saw some renewed interchange, bringing hyaenodont carnivores of the genera *Hyaenodon* and *Pterodon* into North America, and the brontotheres into Europe. At this time, too, the cat-like carnivores (Felidae) made their appearance in both areas.

But the late Eocene interchange was moderate, and in the main the European fauna continued to evolve tranquilly along its own lines, little affected by the transient connexion. At this time the selenodont or crescent-toothed artiodactyls entered upon a great expansion and populated the European scene with numerous, mostly small forms. The most remarkable were the cainotheres with long hind legs not unlike those of a hare. Perhaps they moved with a bounding or hopping gait and filled the same niche as the hares today, although they did not possess gnawing front teeth.

The migration route in the early Eocene was presumably a North Atlantic land bridge, as in the Paleocene. But what about the faunal interchange in the later Eocene? Let us turn to the Eocene history of Asia.

The vast Asiatic land mass formed a block, bounded to the southwest by Tethys, while a great seaway, the Turgai Strait, extended along the eastern side of the Ural Mountains from Tethys to the Arctic Ocean. In the early Eocene, this seaway still formed a complete barrier to direct faunal interchange between Europe and Asia. But the sea gradually retreated enough to open a temporary land connexion, the Tura bridge, in the middle Eocene. At this time the remnants of the interior sea formed a northern and a southern embayment, but the two were soon to unite again, flooding the bridge. As long as it existed, however, terrestrial animals were able to disperse over it.

The climate was temperate to subtropic in northern and eastern Asia, tropical in the south. Fossil floras as far north as the New Siberian Islands have a temperature aspect in spite of their extreme position in the Arctic, and a great fauna of browsing mammals in what is now the Gobi Desert proves that Mongolia was covered by luxuriant forests in the Eocene. A late Eocene flora in Manchuria is intermediate in character between temperate and subtropical and probably marks the boundary between these two types of climate. The boundary apparently stretched from the Tsinling Mountains along the Sungari Valley to the Sikhota Alin range, central Japan and southern Kamchatka. Floras north of this line, for instance from the Ishikari coal field of Hokkaido, from Sakhalin, and from eastern Siberia, have a temperate character. On the other hand, the early Eocene flora at Takashima in Kyushiu consists of a subtropical swamp vegetation with palms (*Sabalites*), figs, sterculia (a relative of the tropical cacao tree) and *Cinnamomum*; this resembles Burmese and Assamese floras of the same date.

Marine faunas with molluscs and nummulites from Japan show the warming effect of the Kuroshio stream to have been notable quite far to the north. In Baluchistan and the Indus region volcanic activity on a gigantic scale was a recurrent phenomenon through much of the early Tertiary, resulting in the creation of the Deccan basalt plateau; the lava beds cover 300,000 square kilometres and reach an aggregate depth of up to 2,000 metres. They intercalate with freshwater beds.

Eocene faunas are known in various parts of this land mass, but the most important come from the area of Mongolia, where a system of fault basins had developed and slowly been filled in with sediments since well back in the Cretaceous. Most of the faunas date from the late Eocene, for instance the famous Irdin Manha about half-way from Kalgan to Urga, and Shara Murun on the Kalgan to Tsagan Nor road. Ulan Buluk in the southern Gobi Desert, however, has yielded a number of mammalian fossils from the early Eocene and these speak as consistently of North American relations to the exclusion of European ones as the Paleocene.

There is, for instance, *Mongolotherium* representing the Dinocerata – a

Figure 15. Last and weirdest of uintatheres, *Gobiatherium* inhabited Asia in the late Eocene. The most conspicuous trait of this large animal is its peculiarly swollen nose region.

group found only in Asia and North America. There is a pantodont representing the last survivors of the Paleocene family Barylambdidae, otherwise known in the New World only. In China, two typical North American tapiroid forms have recently been found in the lower Eocene.

Late Eocene faunas of the Irdin Manha and Shara Murun formations are much richer and also reflect the opening of the Tura route to Europe, if only in a very few immigrants. Otherwise the fauna is Asiatic with some North American elements.

Several forms of brontotheres obviously represent a wave of immigrants from America, perhaps in the middle Eocene, as the late Eocene forms have already diverged to some extent from their New World allies. There is also the carnivorous condylarth *Mesonyx*, an animal about the size of a wolf but with a head that looks somehow too big for the body and the short legs; first found in the Bridger, it now appears in Asia. A further link with

America is seen in the rhinoceros *Amynodon*, also found in the late Eocene of the Uinta. It initiated the amynodont line of aquatic rhinos already described.

In the grotesque *Gobiatherium* of the late Eocene we perhaps see the descendant of early Eocene *Mongolotherium*, a dinocerate with a hornless skull and powerful sabre-like canine teeth. *Gobiatherium* had lost the sabres, the skull was enormously elongated and the nasal part highly expanded, perhaps as a support for a rhinoceros-type keratin 'horn'. Evidently this is the end stage of a very long history of divergent evolution, so it emphasises the separateness of Asia after all.

A bizarre feature in the Irdin Manha is the enormous size reached by some of the creodonts and also the mesonychids (carnivorous condylarths). The biggest of them all, the mesonychid *Andrewsarchus*, was a fantastic brute with a skull a yard long, and related American forms were nearly as large. Many of these big forms probably were herbivorous and, like the bears, found large size an asset in the struggle for existence. Mussel-eating or scavenging habits have also been suggested. The oxyaenid creodonts also tended to large size, culminating in the monstrous *Sarkastodon* of Mongolia, almost as big as *Andrewsarchus*. *Sarkastodon* may be a descendant of the North America *Patriofelis* of the middle Eocene, an animal the size of a bear; that genus, in turn, may be derived from early Eocene *Oxyaena*, a dog-sized form.

The Eocene fauna of southern Asia is not so well known as that of the north. The most important land fauna known at present is that of the Pondaung Formation in central and southern Burma. It rests on marine sediments and consists of estuarine sediments gradually passing into freshwater deposits, indicating a regression of the sea. The fauna is a southern variant of the Irdin Manha theme and is of about the same age (late Eocene). We may note the presence of brontotheres, amynodont rhinos and primitive semi-tapirs; the last-mentioned are distinct from both the European and North American ones.

A special Asiatic contribution was evidently the Anthracotheriidae, which appear in Europe and Asia in the middle Eocene and remained common in Asia throughout the Tertiary. They were fairly large, primitive looking artiodactyls with a four-toed foot and a somewhat pig-like skull and general appearance. Much later, this group probably gave rise to the hippopotami.

As we have seen, the three northern continents had been in contact, at least intermittently, throughout the early Tertiary and so formed the World Continent of the Paleocene and Eocene. The wide expanse of

Tethys still separated Africa from this continent except in the west, where the separation was only effected by the comparatively narrow Betic and Sub-Betic straits. Here, the Iberian Island may have formed a stepping-stone for occasional intermigration.

Africa was still partitioned by high waters in the Eocene. What is now Morocco, Algeria and Tunisia formed a large island; unfortunately all we know about its vertebrate life are the fish and reptiles that inhabited its coastal waters. Africa south of the Sahara was a large continent, which gradually extended northward along the course of the present-day Nile as the land was built up by a predecessor of the modern river. Here, too, the Fayum and Mokattam deposits give less information about the Eocene terrestrial life of Africa than about the marine fauna.

The earliest land fauna of Africa is middle or late Eocene; it comes from Senegal and contains the first known representative of the great order Proboscidea (elephants and mastodonts). There is little doubt that this remarkable group of animals originated in Africa. This early form, called *Moeritherium*, reappears in the late Eocene of Egypt; this later form differs from its Senegalese ancestor only in being somewhat larger. Though a primitive proboscidean, *Moeritherium* is now regarded as an aberrant form not very close to the ancestry of later members of the order. It had a very elongated body, upwards of three metres in length, and was as short-legged as a dachshund in proportion; the small head carried eyes that had shifted far forward towards the nostrils. *Moeritherium* was evidently a highly specialised aquatic form, though its cheek teeth and the tusk-like incisors indicate its proboscidean affinities.

Figure 16. *Moeritherium*, one of the earliest known proboscideans, was about the size of a tapir, and probably amphibious in habits. Its remains have been found in the late Eocene of Senegal and North Africa.

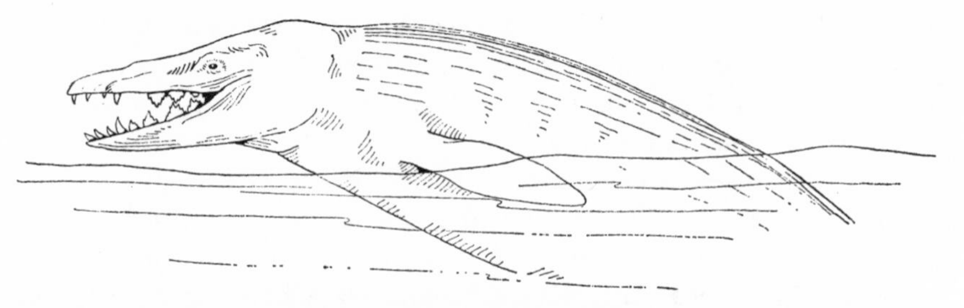

Figure 17. Archaic whale, *Prozeuglodon* of the late Eocene, was as large as a porpoise but still very primitive in many respects. Its serrated cheek teeth may suggest that whale ancestry lay within Creodonta or Carnivora.

Moeritherium shows some resemblance to the Sirenia or sea-cows, which in fact are regarded as an offshoot of the same stock as the proboscideans, and thus also should be of African origin. Sea-cows are, in fact, found for the first time in these Egyptian deposits, and in strata of about the same age in Jamaica; so the Sirenia were widespread at this early date. The early sirenians show the characters of the order in an incipient stage – the peculiar massiveness of the bone structure, the downturning of the snout, and the gradual reduction of the teeth.

Of greater importance are the whales, which make their appearance in the middle Eocene of Egypt; whether this really proves an African origin is doubtful, however, for at least one specimen has also been found in the mid-Eocene of North America, and one fragment from the lower Eocene of England proves the presence of whales in the Atlantic at that early date. In the late Eocene, whales had invaded the seas in large numbers and great variety and are found in marine deposits of North America, Europe, and Africa.

The earliest whales, or Archaeoceti, show a rapid adaptation to a completely aquatic life. The hind legs were reduced to vestiges not projecting from the body, just as in modern whales. The largest of the early whales, the zeuglodonts of the late Eocene (*Basilosaurus*, etc.), were somewhat serpent-like creatures up to fifty-five feet in length and must have resembled certain sea-going lizards of the Cretaceous very closely. (A famous maker of fake exhibitions in the nineteenth century combined the vertebrae of several individuals to create a monster a hundred metres long.) The serpentine body of the archaeocetes is more primitive than the compact body of a modern cretacean. The head was comparatively small and separated from the body by a distinct neck, while the large head of a modern whale is joined to the body without an externally visible neck. The head of the archaeocetes is not unlike that in certain creodonts, but the snout was elongate and the nostrils were half-way up to the top of the head. The front teeth were simple cones, but the cheek teeth were powerful, shearing structures not unlike those in creodonts; this is especially true for the middle Eocene archaeocetes and suggests that the whales arose from some member of the Carnivora or Creodonta which lived on fish, just as the seals were to evolve much later from swimming carnivores.

Eocene history in both North America and Europe is a very long and complicated one, as may be seen from Table 4. It spans eighteen or nineteen million years and is subdivided into four or five ages. The blanks in Asia will probably be filled in before long; faunas of early and middle Eocene date are already known, though not so well as the great late Eocene faunas. In Africa the situation is less promising, but diligent search may still provide information on the land faunas of this continent in the early Tertiary. At present it comes into focus with the Fayum and the transition from the Eocene to the Oligocene.

Again, when we look at the familial diversity of the mammals in the Eocene, the record is one of great increase. The number of families known rises from forty-one in the late Paleocene to ninety-one in the late Eocene. Of course, it should be remembered that part of the seeming increase is really due to incompleteness of the Paleocene and early Eocene record. Still, the main part of the increase is probably real.

If the Paleocene was an epoch of conquest, the Eocene may be regarded as an epoch of consolidation. The mammals had definitely emerged as the dominant group of land animals and were now crowning their victory by

Table 4. The Eocene of the World Continent

Epoch	Africa (faunas)	Asia (faunas)	North America (ages)	Europe (ages)	Date (Million years)
Oligocene	Fayum	(Mongolia)	Chadronian	Sannoisian	
					37
Eocene		Irdin Manha Pondaung	Duchesnean	Ludian	
					40
	Mokattam		Uintan	Bartonian	
					45
			Bridgerian	Lutetian	
					49
		Ulan Bulak	Wasatchian	Ypresian	
				Sparnacian	55
Paleocene		Gashato	Clarkforkian		

Table 5. Number of mammalian families in the Eocene of the World Continent

	Late Paleocene	Eocene		
		Early	Middle	Late
Total number of families	41	51	64	91
First appearances	14	19	25	32
Last appearances	9	12	5	20
Percentage newcomers	34	37	39	35
Percentage extinctions	22	23½	8	22

conquests of the air (bats) and seas (whales, sea-cows). Throughout the Eocene, the origination of new families was continuing at a high rate, though not as high as in the Paleocene. It is true that the percentage of newcomers remains rather stable (varying from 34 to 39 per cent) for each part of the Paleocene and Eocene, but as the Eocene was almost twice as long as the Paleocene, the true time-rate of evolution seems to have abated to some extent. The extremely high rate in the Paleocene, of course, reflects the scramble for completely unoccupied ecological niches in that epoch.

Extinction of archaic forms was comparatively great in the early Eocene, but afterwards there was little extinction on the family level until the late Eocene. At that time, however, several families died out at the transition from the Eocene to the Oligocene. This initiated a weeding-out which was to continue throughout that epoch.

Chapter 4

The Oligocene: epoch of transition

We must extend by long epochs the most liberal estimate that has yet been made of the antiquity of Man.
T. H. HUXLEY: On some fossil remains of man, 1863.

THE OLIGOCENE was still a warm epoch. The temperature of the abyssal waters of the Pacific, which is today less than 2° C, was over 10° C (50° F) in the middle Oligocene. But we know that climates tended to become gradually cooler all through the Tertiary, and this trend was evidently present in the Oligocene too. In an attempt to graph the temperature change of the sea off the Pacific coast of North America, Durham in 1950 showed the probable past position of the February isotherms, representing the coldest temperatures of the year. In this graph the 25° C isotherm moved southward during the Eocene from a position about 48° N (off Seattle) to perhaps 34° N (the latitude of Los Angeles), to continue in the Oligocene to about 25° N (the latitude of Cape Sable, Florida). The last-mentioned is the present-day latitude of the 20° C isotherm. Durham's estimates are based on the distribution of fossil molluscs in the marine deposits of the Pacific coast; they are in good accordance with the paleotemperature measurements of the abyssal water.

In North America, Florida and the Gulf Coast were still under water in the Oligocene; otherwise there is little inundation, except for strips of the Pacific coast. The Mississippi River was continuously feeding new deposits of sand and clay into its estuary, while the bottom of the sea gradually sagged under its increasing load. In fact the total depth of the Cenozoic deposits off the coast of Louisiana is some ten kilometres, and the underlying beds are now flexed down into a geosyncline stretching east and west parallel to the coast.

In eastern North America the sea had retreated from the margin of the continent and the uplift was leading to a gradual re-emergence of the

Appalachians, which had for long been buried under their old peneplane. In the Cordilleran region, on the other hand, levelling had reached a stage when the mountains were almost completely obliterated. The area was now a plain some 2,000–3,000 feet above sea-level, over which the sluggish, meandering rivers spread a blanket of fine-grained deposits. The deposits are comparatively thin but contain an unparalleled fossil record of the herds of mammals that inhabited the area. The abundance and the complete preservation of fossil skeletons make this a land of wonder for the vertebrate paleontologist. (See Plate 4.)

The Oligocene in western North America, then, was a time of comparative quiet after the dying away of the Laramide mountain-building in the Eocene and before the first convulsions of the Cascadian revolution in the Miocene. However, there were numerous active volcanoes in the Oligocene.

The Oligocene of North America is divided into three ages, the Chadronian, Orellan and Whitneyan. The main fossiliferous sequence is that of the White River series of deposits that forms a vast badland area extending over parts of the Dakotas, Nebraska, Colorado and Wyoming. The lower or *Titanotherium* beds date from the Chadronian, while the upper or Brulé beds cover both the Orellan (*Oreodon* beds) and Whitneyan (*Protoceras* beds). The White River series consists of stratified clays with large amounts of volcanic ash and at least some of the beds were probably deposited in temporary lakes – so-called playas – which formed in high-water seasons and evaporated in the dry season.

Of course, the fact that the Oligocene fossil record of North American mammals comes mainly from this area results in a certain bias: we know much about plains animals and little about the fauna of other habitats. In California, however, the upper part of the Sespe series gives some glimpses of Oligocene land mammals of a somewhat different kind, while the deposits of the Gulf coast yield information on the fauna of the sea.

Several fossil floras are known from the Oligocene and confirm the impression of a gradual cooling of the climate compared with the Eocene. A mid-Oligocene flora with a radiometric date of 31·2 million years comes from Goshen-Willamette in western Oregon. Among its species may be noted figs, laurels of several kinds, a southern magnoliacean (*Drimys*), the tropical coastal rosacean *Chrysobalanus*, two species of oak, a holly, soapberry, etc. In the large size of the leaves, the preponderance of simple over compound leaves, and the predominance of pinnate nervations in the leaves, as compared with palmate, the Goshen flora very closely resembles that of present-day Panamá; furthermore, thick leaves constitute about ninety-eight per cent of the total both in the fossil and the recent flora. On the other hand, the Goshen flora differs decisively from the recent in the smaller number of entire-margined leaves and the low frequency of dripping points. It would appear that the Goshen flora lived in a moist but slightly cooler environment than the present-day tropical flora of Panamá.

Another flora from LaPorte on the western slopes of the Sierra Nevada, in northern California, is very close to Goshen in age (date 32·4 million years) but differs in some respects, notably in the presence of palms and cycads. As at Goshen, the laurel family is much in evidence; here it includes several species of the genera *Cinnamomum*, *Laurophyllum*, *Ocotea*,

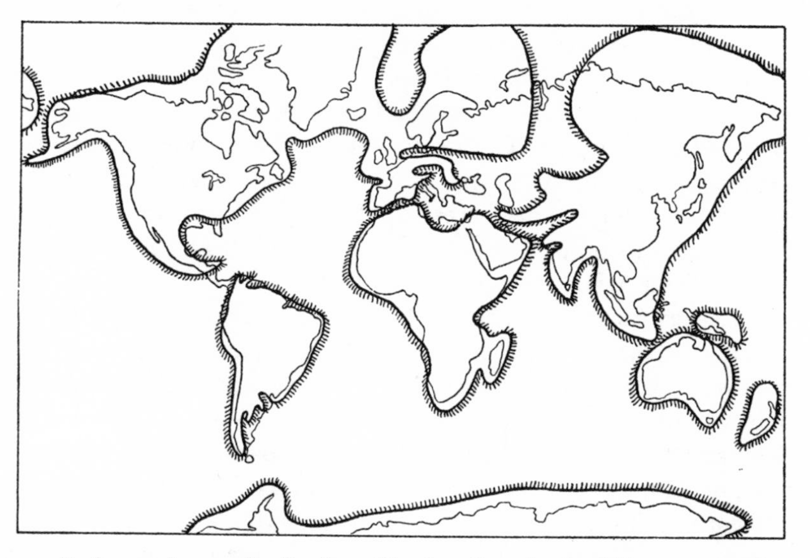

Figure 18. Approximate distribution of land and sea in the Oligocene epoch.

and *Persea*. The lily family is represented by the sarsaparilla bush. Oak, fig, elm and sweet-gum are also present.

A very rich flora comes from Florissant, Colorado, and is thought to be of approximately the same age. It is the local successor of the Green River flora and resembles it in many respects, but has a decidedly less tropical stamp and also suggests somewhat drier conditions. We may note the presence of spruce, several species of oak, poplar, *Carpinus* and others. On the other hand a redwood very close to the Californian form, and a cedar of coastal type, indicate moister patches.

Turning now to Asia, we may note that Mongolia continues to furnish the most important evidence of the land mammals of the Oligocene. A rich fauna from Ardyn Obo correlates with the Chadronian; the Houldjin gravels near Irdin Manha are mid-Oligocene, while Hsanda Gol in the

Tsagan Nor area represents the later Oligocene. Very important late Oligocene faunas have, however, been unearthed in Kazakstan. Other localities from Turkey in the west to China, Korea and Japan in the east give other glimpses of the land mammals in Asia. But the Indian peninsula is a blank and much remains to be discovered about the Oligocene in Asia.

In Europe the three divisions of the Oligocene are termed the Sannoisian, Stampian and Aquitanian. However, the Aquitanian is sometimes included in the Miocene, and the Stampian or middle Oligocene may be divided into an early (Rupelian) and late (Chattian) sub-age.

The Oligocene was the scene of the last great inundation of Europe. Once more, and for the last time so far, the sea spread far and wide across the low-lying lands, dividing Europe into an archipelago. A seaway extended from the North Sea eastward to the Caspian basin, flooding the Low Countries, northern Germany, Poland and a wide expanse of the Russian plain. To the south this interior sea, which is called the Lattorf Sea, connected with Tethys across the Black Sea basin. In this way central and western Europe were completely severed from the Scandinavian shield and from Asia.

The Lattorf Sea overran the forests of the previous age in the Baltic region, and of the decaying logs finally nothing but the drops and lumps of resin were preserved. This is now Baltic amber and pieces of amber may contain various objects once trapped in the resin – twigs, leaves, insects, spiders, and so forth. The insect fauna of the Baltic amber is one of the richest known to us. In its composition it appears quite modern; there are even some species that seem to be identical with species that are still in existence today. We can only conclude that the insects have evolved at a much slower rate than, for instance, the mammals; for all of the mammalian species of the Oligocene are quite different from any species living today.

Where the Eocene forests had grown, there now lived the bottom fauna of the Lattorf Sea. Oysters would inhabit the hard bottoms, together with the spiny bivalve *Spondylus*; on sandy bottoms the little *Crassatella* would nestle in the depressions. The area was populated by many molluscs that are now typical of the subtropical and tropical seas, such as *Fusus*, the beautiful and poisonous *Conus*, and many others. But the big sea snakes that had been so common off the European coast in the Eocene were gone.

Further west the Paris basin was inundated for the last time. The shallow waters in the inner part of the basin were brackish or fresh, and have left a deposit of freshwater limestone with a characteristic mollusc fauna. Brittany continued as a dry highland, at times almost isolated from the remainder of Europe.

Central Europe, between the Lattorf Sea and the Flysch trench north of the gradually emerging Alpine islands, formed an elongated land barrier from France in the west to Crimea in the east. Another large island was formed by the Scandinavian and Russian platform, and Britain also was an island, completely emergent except for the Hampshire basin.

Along the course of the present-day Rhine a depression was now forming straight across the Central European island. This depression, the Rhine Graben, was thrown down as a block sliding along fault lines in relation to the blocks on either side. The area now became a string of lakes and was later invaded by sea water.

In the middle Oligocene the Lattorf Sea tended to retreat to the west, but also encroached southward to some extent. In the late Stampian only an embayment remained, extending to the Oder in the east. Further west a brief mid-Oligocene transgression in the Paris basin was followed by a regression and the basin now formed a freshwater lake, never again to be invaded by the sea.

Meanwhile, dramatic developments were going on in the Rhine Graben and along the Tethyan coast. The sea again invaded parts of the graben in the middle Oligocene; volcanic activity became intense in the late Stampian. Further south and west, the Apennines and Pyrenees had already undergone major folding and uplift during the Eocene-Oligocene transition: now it was the turn of the Alps themselves. As the suction in the mantle of the earth pressed the African and Eurasian blocks towards each other, the main or Helvetian phase of the Alpine orogeny was enacted. As the great mountain range developed, its flanking trench was gradually pushed forward; the old Flysch trench became incorporated in the rising masses and instead a new type of sediment, the so-called molasse, was deposited in the shoaling trench. In the south, near the slopes of the mountains, the molasse takes the form of Nagelfluh, a conglomerate ('puddingstone'); further north it consists of sandstones and finally, at a greater distance, of clayey sediments. Many of the molasse sediments have preserved a record of the land fauna.

In the Aquitanian stage the sea retreated still further and most of Europe was emergent. The only remnant of the one-time Lattorf Sea was now an embayment covering Jutland. The molasse trench at the foot of the Alps was above sea-level and wide, shallow lakes were forming within it and in other areas in the Rhine Graben and in central and southern France.

Although temperatures were gradually being lowered, Europe still had a warm climate. Sabal palms grew along the coast of the Lattorf Sea, lignite formation continued in Germany, and crocodilians flourished. Still, the climate was evidently cooler than in the Eocene; there is an increase of conifers and deciduous trees. Locally, dry conditions are indicated, for instance in the southern part of the Rhine Graben where deposition of rock salt alternated with incursions of the sea.

Early Oligocene land fauna is especially well known in France. Not only are part of the Quercy phosphorites of this age (alas, the Eocene and Oligocene material is almost hopelessly mixed in the old collections!), but there are also the freshwater marls of Ronzon at Le Puy with a rich fish, bird, reptile and mammal fauna; the Marnes supra-gypseuses and the Brie limestones in the Paris basin; and the fauna from Lobsann in the Rhine Graben, as well as many others. In England this stage is represented by the mammalian fauna of the Hampstead beds east of Yarmouth.

Early Stampian (Rupelian) localities are more widely scattered, from the Fontainebleu sands of the Paris basin to the rich fish faunas in Hungary, mammals of the Apennine island in Italy and the molasses of Lérida in Spain, to mention only a few. The late Stampian (Chattian) is a blank in the Paris basin but well represented in the other areas, both in France, the Rhine Graben, and in eastern and southern Europe. Finally the Aquitanian is represented by the famous mammalian fauna from St Gérand-le-Puy in Allier, from Weisenau and Mombach in the Rhine Graben, and from numerous molasse localities in Switzerland and elsewhere north of the Alps.

A rich record of the later Oligocene life in Europe is found at the village of Rott on the northern flank of the Siebengebirge in northwest Germany. Here there are deposits of thinly laminated lignites from which oil was formerly obtained; active mining was abandoned long ago, but the site has yielded a rich harvest of fossils, amounting to some 300 species of plants and 700 of animals. The sediments were formed in fens with rank stagnant waters, in which the lack of oxygen contributed to the preservation of organic remains. The rich diatom flora of the pools has in some cases impregnated the sediments with silicic acid. The lamination is a result of seasonal variation; there are about sixty-five of these laminae to a centimetre. It is thought that the area suffered two annual drought periods, in which the water evaporated, and two rainy seasons with replenishment of the pools.

The lakes were surrounded by woods with poplar, willow, elm, sweet gale (*Myrica*), walnut, hickory and occasional sabal palms, thickly interwoven with lianas to form a dense, impenetrable jungle. This suggestion is in rapport with the scarcity of solitary types of bees among the insect fossils (they prefer open ground) and the great number of social forms (which make their nests in hollow trees). At a distance from the swamps, the forest was somewhat more open and consisted of hornbeam, sycamore, beech, soapberry tree, sweetleaf (*Symplocos*), *Catalpa* tree and others. The slopes of the Siebengebirge, towering over the fen landscape, would carry oak, laurel, dogwood, heath, sapodilla, oleander and numerous other plants. The great majority of the fossil animals are insects and the lakeside woods would be infested with mosquitos and flies of all kinds. Dragonflies, beetles, wasps and bees, cicadas and various other forms, totalling about 650 species, are also found. The lakes were inhabited by fish, crocodilians, salamanders, and frogs; sparse remains of land mammals are also found. The picture is that of a subtropical country with a highly developed seasonal variation in humidity and a great range of habitats from high hills to swamp.

All over the continents of Eurasia and North America the mammalian fauna of the early Oligocene was rather homogeneous. Evidently the regression of the late Eocene had provided dry land for the expansion and migration of land animals from continent to continent, and in the north the climate was presumably still mild enough to permit a varied fauna of land mammals to cross the Bering bridge. As the epoch wore on, however, divergence between the Old World and the New became gradually more marked, as if there had been a break in the land connection, or perhaps a climatic deterioration in the Arctic region, making the bridge inaccessible to certain types of animals.

At the same time another kind of change may be sensed in the mammalian faunas. It is as if the resistance of the environment started to make itself felt: the number of families that had been steadily growing throughout the Paleocene and Eocene was now checked and reduced for the first time. This may be connected, too, with the gradual lowering of the temperature. Perhaps the climate shift led across a kind of threshold where the effects of the cooling suddenly broke through and were strongly reflected in the organic world. Or perhaps it was a phase inherent in the pattern of adaptive radiation. In the Paleocene and Eocene numerous different types of animals had evolved to fill the diverse ecological niches. Unavoidably, many forms of different ancestry radiated into analogous niches and sooner or later were thrown into competition. Out of this emerged as victors the most efficient types, while the less well-adapted forms tended to wane and perhaps to become extinct.

In fact there was a gradual weeding out of many forms. In the beginning of the Oligocene the fauna of the northern continents was more varied than at any previous time, with some ninety-five families of mammals. During the epoch, however, more than one-third of them became extinct, and at the same time the production of new families abated. It was an epoch of transition: archaic forms died out and the modern families and orders were taking over.

The multituberculates were gone, and several families of archaic insectivores died out at the end of the Eocene or during the Oligocene; among them may be noted the otter- or desman-like pantolestids and the rodent-like insectivores. Instead, modern insectivores – hedgehogs, shrews, and moles – became numerous, both in the Old World and the New (hedgehogs survived in America to Pliocene times). The ancestry of the hedgehogs may be traced back to the late Cretaceous. A family related to the hedgehogs are the Dimylidae, which arose in Europe in the Oligocene, flourished there in the Miocene, but became extinct in the Pliocene without ever reaching any other continent as far as we know.

The taeniodonts, the tillodonts, the uintatheres, and the rare northern notoungulates became extinct before the beginning of the Oligocene. The same fate befell almost all the condylarths except for a few relict populations in the Old World (and in South America). A relict of the coryphodonts also persisted in East Asia to the middle Oligocene, or some twenty million years after the extinction of this group in Europe and North America.

We also lose sight of almost all the prosimian primates that were so common in the northern continents in Paleocene and Eocene times. With the transition from the Eocene to the Oligocene there is a coincident shift of the centre of primate distribution to the southern continents. At about this time, some of the North American prosimians were evidently rafted to South America, to give rise to the South America monkey stock that makes its first appearance in Oligocene strata. African Paleocene and Eocene primates are so far unknown, but in the Oligocene Africa emerges as a major centre of primate evolution. Unfortunately, the Oligocene in Asia is at present a blank as far as primates are concerned.

But extinction in the north was not complete, at least not in North America, where a few prosimians persisted in the Oligocene and one form has even been found in early Miocene deposits, a discovery as startling as the generic name (*Ekgmowechashala*) given to this defenceless creature. All these survivors belong to the tarsier-like prosimians (family Omomyidae) which by some authors are regarded as ancestral to the South American monkeys. Others favour a derivation from the lemur-like Adapidae, which became extinct in North America well before the end of the Eocene.

The decline of the condylarths and other primitive ungulates was obviously connected with the deployment of the modern ungulate orders –

the perissodactyls and artiodactyls. But again within these orders there was a weeding out of ancient families, which were crowded out by more progressive forms. The horse-like palaeotheres of Europe survived into the earliest Oligocene, then died out. The big brontotheres with their V- or Y-shaped bone protuberances on the nose were at the culmination of their career in the Eocene-Oligocene transition, yet they were all gone in North America in the middle Oligocene, and in Asia in the late Oligocene. This family never amounted to much in Europe, where a few rare forms appeared in the early Oligocene, only to become extinct shortly afterward.

Instead, Europe may be thought of as rhinoceros land in the Oligocene; although present and varied in all the northern continents, they had fewer rivals in size in Europe than elsewhere. The amynodonts or amphibious rhinos were especially common in Asia, where they lasted into Miocene times, many million years after their extinction in Europe and North America; a parallel to the history of the amphibious coryphodonts. But the true rhinoceroses flourished mightily, reaching their apogee in the late Oligocene of the Old World with more than fifteen genera.

Among the true rhinos may be noted the peculiar *Diceratherium* with a transverse pair of horns on the nose, a long-lived and successful group, present both in the Old World and the New. In Eurasia, the medium-sized, hornless *Aceratherium* was another common and long-lived form that persisted into the Pliocene. However, the prize for success and longevity must nevertheless go to the genus *Dicerorhinus* which made its first appearance in the Rupelian (early middle Oligocene) and still exists today in the Sumatra rhino, after having produced many of the distinctive rhinoceroses of Eurasia in the Oligocene, Miocene, Pliocene, and Pleistocene. *Dicerorhinus* is thus one of the oldest genera of mammals in existence, perhaps only exceeded by one or two genera of bats such as *Rhinolophus* (horseshoe bats) which may date back to the late Eocene.

The giant rhinos or indricotheres were another interesting group, relatively short-lived but of awesome size. The biggest stood more than five metres high at the shoulder. With their long necks and relatively small, hornless heads they deviated considerably from the normal rhinoceros type; their ecological role probably resembled that of the present-day giraffes. As far as is known the indricotheres were the largest land mammals of all times. (See Figure 19.)

The horse family, after its extinction in Europe, continued its existence in North America with the Oligocene genera *Miohippus* and *Mesohippus* – three-toed forest horses in which the premolars had become completely

Figure 19. A treetop browser much like the giraffe, the hornless giant rhino *Indricotherium* attained a shoulder height of seventeen or eighteen feet and was the largest and heaviest land mammal of all times. A mainly Asiatic group, the indricotheres were in existence from the early Oligocene to the early Miocene.

molar-like, but the cheek teeth still were low-crowned and suited for leaf-eating habits. There was some tendency to a size increase. On the whole, however, equids are not in the foreground in the Oligocene, and the same holds for the tapir-like perissodactyls, most of which became extinct at an early date. Only the tapir family itself survived in the long run.

The most conspicuous of the Oligocene artiodactyls were apparently the 'giant pigs' or entelodonts. They were very large, grotesque-looking animals with peculiar bony protuberances jutting out from the cheek region of the enormous head; their function is difficult to imagine. The entelodonts probably were somewhat pig-like in appearance but their limbs were long and they were probably fairly fleet of foot despite their ungainly size.

True pigs were already present in Europe in the Oligocene, and peccaries in North America. Among the smaller artiodactyls, the hopping cainotheres continued to flourish in Europe during the Oligocene, whereas the

Figure 20. *Archaeotherium*, an entelodont or 'giant pig' from the Oligocene of North America and East Asia. Bony protuberances of cheek and jaw may have been used in fighting. The size of a large cow, *Archaeotherium* was a long-limbed and probably fast-running animal.

Figure 21. *Cainotherium* and its relatives, only known from Europe, were small, superficially rabbit-like ruminants; they range from the late Eocene to the Miocene but their heyday is in the Oligocene. This animal was about a foot in length.

numerous primitive families that predominated in the earlier Eocene now dwindled and became extinct.

For sheer numbers the predominant artiodactyls of the Old World were the anthracotheres, which arose in the Eocene and became very common in the Oligocene. Some of these medium-sized, pig-like animals populated North America in the early Oligocene and managed to keep a foothold in that continent well into the Miocene. But they never became numerous, evidently because the 'anthracothere niche' in North America was securely held by a quite different group of artiodactyls, the oreodonts.

The oreodonts are an exclusively American group, which was prolific in the Oligocene and Miocene. Unlike the anthracotheres, they belonged to the true ruminants with crescentic cheek teeth and have their closest relatives in the camel family. In spite of this relationship their exterior was probably often somewhat pig-like with a long body and short limbs; they are sometimes referred to as 'ruminating swine'. True camelids were also present in North America since the late Eocene.

There are also the beginnings of the higher ruminants or pecorans, in the form of usually small animals related to the present-day chevrotains.

Figure 22. Most wide-ranging of anthracotheres, *Bothriodon* was distributed in North America, Europe, Asia and Africa in Oligocene times. Length about five feet.

The Oligocene pecorans are grouped into four families with a somewhat uneven distribution over the northern continents. Most of them were hornless, but in the American family Protoceratidae there was a tendency to develop horn-like bony protuberances on the skull, sometimes of very bizarre appearance.

The carnivorous forms which preyed on these plant-eaters were also undergoing a transition. The creodonts, or primitive carnivores, declined sharply at the end of the Eocene. A few forms survived in Oligocene times; in Asia a member of the Hyaenodontidae survived as late as the Pliocene, and other members of the same family have been found in Africa in Miocene deposits. In general, however, the carnivore niche was invaded massively by species belonging to the modern families.

Four families – the viverrids, cats, mustelids, and dogs – had appeared in the late Eocene and now proliferated greatly. The viverrids (civets, genets and related forms) have always been restricted to the Old World, but the closely related cat family established itself in both hemispheres. Some of the Oligocene cats were ponderous, heavily-built, sabre-toothed forms that evidently preyed on slow-footed large game. Others were slim,

Figure 23. The redoubtable *Eusmilus* was the largest of the sabre-toothed felids of the Oligocene; it was about the size of a leopard, but more heavily built. *Eusmilus* was present in both the Old World and the New.

dashing sabre-tooths; while others again had normal cat-like teeth used for biting and not, as in the case of the sabre-tooths, for stabbing. One skull of the leopard-sized *Nimravus*, belonging to the last-mentioned 'normal' group, and found in the White River deposits, tells a remarkable story. It shows in the left frontal region a terrible but partly healed wound, probably inflicted by the sabre of the formidable *Eusmilus*, one of the largest and most powerful sabre-toothed forms known. The sabre did not penetrate to the brain and so the unfortunate *Nimravus* survived long enough for the partial healing of the wound.

Figure 24. The Oligocene otter *Potamotherium*, a very advanced water-living form which may be close to the ancestry of the fur seals. Fossils are especially plentiful in parts of France.

Mustelids were still comparatively rare in the Oligocene, but they already included – besides ordinary marten-like forms – such highly specialised animals as the otter *Potamotherium*. This was an aquatic form, in some respects more highly adapted to swimming than the present-day otter *Lutra*. The dog family, Canidae, on the other hand was highly abundant and varied. It had a great size range from small animals the size of a fox to large, somewhat bear-like forms like the big *Amphicyon* that invaded all

Figure 25. In the large, somewhat bear-like dogs of the *Amphicyon* type, the tubercular back teeth were greatly developed, suggesting predominantly herbivorous habits. They were an eminently successful, long-lived and widely distributed group.

the northern continents and continued its successful career well into the Pliocene.

The early history of the true wolves and dogs seems to be confined to North America and goes back to the Oligocene *Mesocyon*, which is more primitive than modern dogs, with a long body and tail, and limbs less adapted for running.

As an offshoot of the true canids, there also appear some forms which seem to represent an early stage in the evolution of the true bears, although the Oligocene representatives were still of moderate size. Another group of small, civet-like dogs in the Oligocene may be close to the ancestry of the raccoon family or Procyonidae. This family makes its appearance both in Europe and North America in the early Oligocene with the same genus (*Plesictis*); presumably it was also present in Asia though it has not so far been identified there. The family enters the record in Asia in the Miocene.

Rodents are common in the Oligocene in both hemispheres. Besides the primitive Eocene groups, many modern families are now present. In the squirrel family may be noted the appearance of the living squirrel genus *Sciurus* in the late Oligocene; so that, oddly, the common squirrel may be regarded as a 'living fossil' almost on a par with the Sumatra rhinoceros. Early forms of the hamster and vole family are also found, and dormice were not uncommon in Europe in the Oligocene. There are also pocket gophers (in America), bamboo rats and birch mice (in Eurasia), beavers and many others. Among the lagomorphs, both pikas and hares are represented in Asia and North America; pikas occur also in Europe but the hare family is not found there.

Such, in brief survey, was the Oligocene mammalian population of the northern continents. In the Oligocene, however, we have to add another continent to the group of interconnecting land areas that we have termed the World Continent. This is Africa, which now at last enters the record with a rich land fauna, showing clear evidence of preliminary contact with Europe and Asia, although at the same time its animal life remains stamped by its long independent history.

In the Paleocene and Eocene, northern Africa was flooded by Tethys. Primitive whales, dugongs, sea serpents, giant sharks and crocodilians were swimming in the big estuary of the primeval Nile. But gradually it silted-up; the ancestral Nile was forever carrying and depositing immense amounts of sediments and pushing the coastline northward. In the Oligocene, the coast was already north of the present-day Fayum area, some sixty miles southwest of Cairo. (See Plate 3.)

In this area there is now the 1,600-feet escarpment of the Jebel Qatrani, looking south over the brackish Lake Qarun, in the midst of a wind-assailed desert landscape. In Oligocene times this was a green delta with forests and savannas and a rich and strange animal life; but now there remain only the fossils, which are found in the exposures of the Jebel Qatrani as the unceasing wind erosion chisels them out of the rock.

The lower 700 feet or so are Eocene in age and bear the fauna with *Moeritherium* and primitive whales described in the previous chapter. This is followed by a long series of Oligocene deposits capped by a basaltic lava flow, which has been dated radiometrically at twenty-five–twenty-seven million years. Since the Oligocene is considered to have ended at about that date, it seems that the entire epoch may be spanned by the upper part of the Fayum series.

Most of the fossils are concentrated in two forest bed zones in which big tree trunks up to 100 feet long are found lying where they were once deposited by the primeval Nile. The lower of these fossil wood zones is in a position suggesting an earlier Oligocene date, while the upper zone may date from the earliest part of the late Oligocene. Some fossils have also been found about midway between these two levels, and so probably are mid-Oligocene in age.

Apart from the rich mammalian fauna, reptiles and birds are also found in the Fayum, an interesting bird fossil being that of an early relative of the extinct elephant birds of Madagascar. There are also giant land tortoises, pythons and crocodilians, both 'normal' types and the extremely long-snouted, gavial-like *Tomistoma.*

The mammals may be divided into immigrants (allochtonous forms) and local (autochtonous) products. Among the immigrants may be noted the anthracotheres, which were quite varied at Fayum. There was also a member of the Cebochoeridae, a family of small pig- or entelodont-like forms which did not, however, reach the size of the entelodonts, or develop such an elongated skull. The family is known from the Eocene and early Oligocene of Eurasia. Rodents are present in the form of an exclusively African family, now long extinct but perhaps ancestral to such living African forms as the cane 'rat' and rock 'rat'.

There are no true carnivores, but the hyaenodonts flourished here with several forms at a time when creodonts were becoming extinct almost everywhere else. There may even be a member of the mesonychid family (carnivorous condylarths). It is evident that Africa may have played the same role as a refuge for archaic forms in the Oligocene, as it does now.

The presence of these Eurasian groups shows that some intermigration was possible. Still, the absence of the great majority of the Eurasian forms, and the presence of many autochtonous African mammals which are not found elsewhere, proves that such migration was difficult and chancy.

The main autochtonous group of Africa are the so-called subungulates – a group of mammalian orders clearly related to each other and evidently of African origin (although recent discoveries in Mongolia indicate that some members of the group succeeded in making the long journey thither at some time in the earliest Tertiary). Here belongs the largest land mammal of the Fayum, a gigantic four-horned creature called *Arsinoitherium,* the sole known representative of its order. In the flesh it may have looked a little like a titanothere, though the two diverging horns on its nose were much more massive and there was an additional, smaller pair over the eyes. Its dentition is very primitive and indicates leaf-eating habits.

The same is true for the conies or dassies, which are the present-day survivors of the subungulate order Hyracoidea. In the Oligocene it was a highly flourishing group, some members of which were as large as a tapir. These giant hyraxes presumably formed large herds and are the most common mammals found as fossils in the Fayum. Like the living conies, the Oligocene forms were probably browsers, for their dentition is closely similar to that of the living.

The sea-cows and proboscideans are other subungulates present in the Fayum. Of the primitive proboscideans from the late Eocene, *Moeritherium* survived in the early Oligocene, but in addition the first real mastodonts are now on the scene. They were of rather modest size, and although one

Figure 26. The first mastodont, *Palaeomastodon* of the Fayum Oligocene, was also one of the smallest, measuring only four or five feet at the shoulders. Long jaws and short trunk were typical of many early mastodonts.

species attained almost the size of a small modern elephant, they did not rival *Arsinoitherium* as the largest mammal in the area. All of the Oligocene mastodonts may be referred to the genus *Palaeomastodon,* characterised by the presence of one pair of relatively short tusks both in the upper and lower jaws. The upper tusks protruded somewhat to the sides and were probably used as weapons, whereas the lower formed a peculiar spoon- or scoop-like structure which may have been used for digging. Very probably there was a long prehensile upper lip in conjunction with a short trunk like that of a tapir.

Finally we come to the primates of Fayum. In a long perspective, they must of course be regarded as immigrants; but they evolved, evidently in Africa, from the prosimian level into monkeys and apes; so that these families, and probably the human family as well, apparently arose in Africa. At the Fayum we see recorded a very remarkable radiation – divergent evolution – of several stocks of higher primates; there seem to be here the beginnings of some four or five different trends, including the one that was to lead to man.

The monkeys of the Old World may well have descended from a creature like tiny *Parapithecus*, a creature hardly larger than a tarsier or squirrel monkey. In common with many prosimians it still had three premolars in each half of the jaw – all the Old World monkeys, apes and man have only two. Its dental characters may indicate that it was transitional between early prosimians and the modern Old World monkeys – the macaques, baboons, langurs and others.

Closely allied to *Parapithecus* is another very small form, *Apidium*, whose remains outnumber those of all other primates at Fayum. There is some suggestion that this creature may have been ancestral to a line of peculiar ape-like primates which we shall meet many million years later in *Oreopithecus* of Tuscany.

There is also a small form which shows some resemblance to the gibbons and may be close to their ancestry. But the best-known primates at Fayum are ancestral to the great apes. Im the early Oligocene there is a small, incompletely known form called *Oligopithecus*, but the late Oligocene *Aegyptopithecus* is recorded on the basis of several jaws and a well-nigh complete skull – one of the two fossil ape skulls known to date. This form was the largest primate of Fayum and attained the size of a gibbon. Although the dentition of *Aegyptopithecus* is clearly that of a true ape, its head was fairly primitive and monkey-like rather than ape-like, with a long narrow snout; also, it still had a tail.

Finally we come to the enigmatic little creature that has received the jaw-breaking name *Propliopithecus* (in the mistaken belief that it was ancestral to the late Tertiary *Pliopithecus*, a gibbon-like form). So far, only a few lower jaws and isolated teeth have been found, but they suffice to demonstrate some very remarkable traits.

One of the most important differences between the teeth in man and in apes is the large size of the canine teeth in the latter, and the development of the foremost lower premolar into a large, pointed, one-cusped structure, elongated fore and aft and with a cutting front edge. In humans this tooth is small and two-cusped, and the canine teeth are small. In both these characters *Propliopithecus* is manlike rather than ape-like. There are also other traits (in the relative size of the molars, for instance) which suggest hominid affinities.

It may also be significant that all the specimens of *Propliopithecus* apparently come out of the mid-Oligocene deposits at Fayum, in contrast with the true apes *Oligopithecus* and *Aegyptopithecus*. These come from the lower and upper forest beds respectively, while the man-like form comes from an intermediate layer without fossil trees. There is here a suggestion that *Propliopithecus* may have inhabited a different type of country, perhaps not so densely forested as the habitat of the true apes and monkeys, perhaps even a savanna environment.

The Oligocene then was an epoch of transition in the primate world: the prosimians were leaving the scene and the first monkeys and apes were appearing, perhaps even the first man-like primates. Again, when we turn our gaze to the seas, we note some evidence of a major transition among the cetaceans. Unfortunately the Oligocene record of the Cetacea is very poor, whereas a great number of marine mammal-bearing beds are known from the Miocene. In these latter, there is a veritable burst of new forms, which obviously must have evolved in the Oligocene. But at present we only have a few scattered finds. From these it appears that the serpent-like archaeocetes, which had flourished in the Eocene, were on the wane. In the early Miocene there persisted only one family of this group, the relatively small, short-bodied dorudonts. On the other hand, precursors to both the toothed whales and the whalebone whales had appeared in the late Eocene, and a few Oligocene finds show that true representatives of both groups were present. Perhaps one day the whole story will be told.

In retrospect, then, the Oligocene may be seen as a comparatively short

Table 6. The Oligocene of the World Continent

Epoch	Africa (Fayum) zones)	Asia (faunas)	North America (ages)	Europe (ages)	Date (Million years)
Miocene		Bugti	Arikareean	Burdigalian	
					25
Oligocene	Upper Fossil Wood Zone	Hsanda Gol	Whitneyan	Aquitanian	
					27
				Chattian	
					29
	Propliopithecus Zone	Houldjin	Orellan	Rupelian	
					32
	Lower Fossil Wood Zone	Ulan Gochu	Chadronian	Sannoisian	
					37
Eocene	Mokattam	Irdin Manha Pondaung	Duchesnean	Ludian	

Table 7. Number of mammalian families in the Oligocene of the World Continent

	Late Eocene	Oligocene		
		Early	Middle	Late
Total number of families	91	95	82	77
First appearances	32	24	7	5
Last appearances	20	20	10	8
Percentage newcomers	35	25	8½	6½
Percentage extinctions	22	21	12	10½

epoch – it lasted some ten or twelve million years only – in which the progress in evolutionary diversity was finally checked. For the first time extinction out-paced origination, and even though the entrance of Africa into the fossil record swells the initial rank of families, the total of land mammal families known in the World Continent decreased steadily from ninety-five to seventy-seven. The main part of the turnover falls in the early Oligocene, and the rate of change abated very clearly in the later half of the epoch. We have come to the end of the Oligocene, and of the Paleogene or early Tertiary as well. It is a more familiar world we are entering now: the world of the Miocene.

Chapter 5

The Miocene: epoch of revolutions

Behold, the Lord maketh the earth empty,
and maketh it waste, and turneth it upside down,
and scattereth abroad the inhabitants thereof.

Isaiah, 24:1.

THE UNENDING succession of new forms of life is in itself a wonder to contemplate: the fact that the earth has ceaselessly produced new types of animals and plants, in a continuously changing landscape, for so many million years. In this perpetual state of flux, when we start from the beginning of the Age of Mammals and approach our own time, there comes a moment when sudden recognition dawns; when for the first time we find a system of land forms that has a familiar look, with creatures that are obviously akin to those of the present day. Perhaps this line could be drawn somewhere near the transition from the Oligocene to the Miocene. At this time the archaic forms of mammalian life were already drastically reduced, and it has been estimated that only one-fourth of the mammal families then in existence have since become extinct. At the same time this was the epoch in which the ancient Tethys Sea was finally broken up by the appearance of the young mountain ranges, and the modern Mediterranean was born. Africa was brought into communication with the Eurasian land mass. Finally, there are the first suggestions of the cold breath of inland glaciation.

When all this is added up, the beginning of the Miocene may be seen as the initiation of new trends in earth history, in quite a different manner from, for instance, the Mio-Pliocene transition which mostly involves just a continuation of these same trends. In fact the Mio-Pliocene transition is quite a bone of contention and much of what is here regarded as Pliocene has been included in the Miocene by various authorities. Even with the minimal scope given here, the Miocene was one of the longest epochs of the Tertiary, lasting about fifteen million years; it is exceeded in length only by the Eocene.

Temperatures continued to fall in the Miocene. In Europe the annual mean temperatures estimated for the late Oligocene flora and fauna of Rott (described in the previous chapter) is 18° C (64° F). An analogous estimate for the mid-Miocene of Oeningen at Lake Constance is 16° C, and for the Pliocene of Frankfurt 14° C. The present-day average in all these areas is about 9° C (48° F). In spite of the lowering of temperature, the Miocene would still appear a hothouse world to us.

It is the more remarkable that there are definite suggestions of local

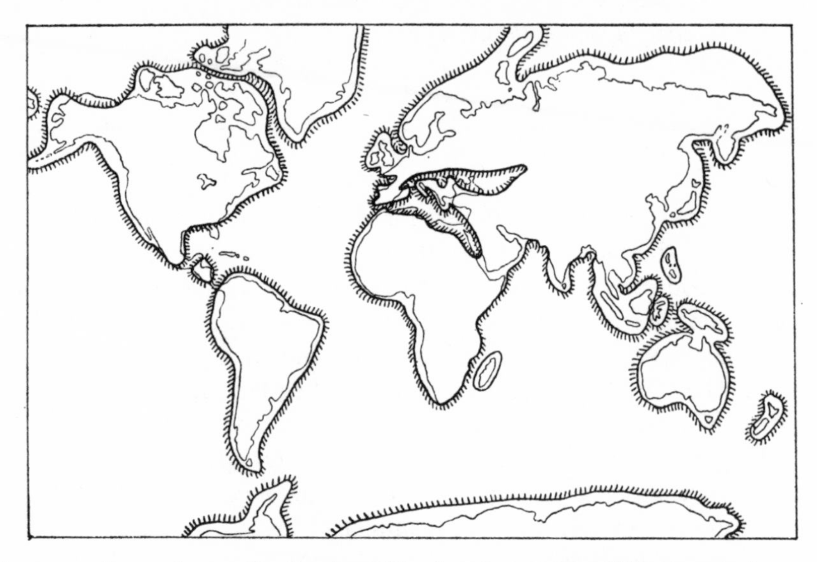

Figure 27. Approximate distribution of land and sea in the Miocene epoch.

glaciation in some parts of the world, for instance Iceland. Here it should be noted that the Gulf Stream, which has a warming effect on Iceland, probably did not exist in its present form in the Miocene. A study of sea paleotemperatures in the Atlantic off Florida in the Oligocene and Miocene gave the surprising result that Oligocene temperatures were about the same as those of the present, and the Miocene some 4° lower. Now at this time there was free communication between the Atlantic and the Pacific Oceans by way of the Bolívar Trench at the point where the narrow Central American isthmus joins South America. It follows that the Equatorial Current, which is now forced to make a U-turn and issues through the Straits of Florida, might then at least in part have drained through the Bolívar Trench. The warming effect of the Florida Current would be lost and there might even be some effect of the cold Labrador Current moving southward along the east coast of North America.

The temperature curves themselves also have some puzzling features. The Oligocene record is comparatively uniform, but in the Miocene there is a suggestion of secular fluctuation reminiscent of that found in the Pleistocene. This is in agreement with recent discoveries of prominent climatic oscillations recorded by changes in the European pollen flora in the Mio-Pliocene. Perhaps this is a result of local glaciation in the Arctic and Antarctic regions, with alternation between glacial and interglacial phases.

A wholly different line of reasoning also leads to the suggestion of polar cold in Miocene times. This is the gradual cessation of intermigration across the Bering bridge of some types of mammals that appear to have a mainly tropical adaptation. From now on the forms exchanged between Asia and North America are mainly or solely such that may be assumed to withstand fairly cold conditions.

Very likely all this reflects the effect of the enormous mountain-building revolutions which were initiated by the Helvetic phase of the Alps in Europe in the Oligocene, and which in the Miocene spread throughout the Tethyan system of Asia, also producing the Cascadian orogeny in North America. The emergence of the new mountain ranges had a profound effect on the circulation pattern of the air masses. Furthermore, with the expulsion of the sea from most of the wide expanse of Tethys, the earth's retentivity of solar radiation was strongly affected. Water holds heat much better than land, from which it is reradiated into space.

In North America the Cascadian revolution of the Miocene initiated a series of mountain-building phases which went on into the Pleistocene and comprised not only the Cascade Mountains themselves (in the states of Oregon and Washington) but a vastly wider area. Mostly it took the form of vertical uplift or downthrow along fault lines, but folding and over-thrusting was also involved in the Coast Range and the Puget Sound area, Washington. Further south, the Sierra Nevada still formed a chain of hills of moderate height, and the Miocene floras indicate that the vegetation and climate was more or less uniform. The western slopes were well watered, but locally the eastern lee brought aridity to the land; there are some basins in Nevada showing evaporation and deposition of rock salt. The main part of the Great Basin supported hardwood deciduous and conifer forests and may have received an annual rainfall of between 900 and 1,300 millimetres spread uniformly over the year. Further south much drier conditions are indicated in the Mojave basin with a live oak, chaparral and thorn forest vegetation. Along the Californian coast marine embayments continued to encroach upon the land.

Volcanism occurred on an unprecedented scale, leading for instance to the formation of the Columbia Plateau with some 5,000 feet of lava beds covering an area of more than 200,000 square miles in Washington, Oregon, and Idaho. A dramatic episode has been highlighted by an unusual discovery in the Lower Grand Coulee basalts of east-central Washington. Apparently the carcass of a *Diceratherium* rhinoceros, drifting in a lake, was here engulfed in lava to become preserved as an impression in the basalt. Now lava of this type is fluid only above 900° C, which

Figure 28. Preserved as a natural cast in the Grand Coulee basalt, Washington: the small rhinoceros *Diceratherium*. This animal, which was less than two metres long, had a transverse pair of nose horns; its history goes back to the Oligocene.

makes it seem odd that the rhino was not simply destroyed by the heat. But it appears that the lava formed a set of squashy 'pillows' with a slightly cooled outer surface and a fluid interior as it flowed into the water of the lake and embedded the dead body. Within the mould, remains of bones were found, which although much warped by the heat made it possible to identify the animal. It must have been a great surprise to the original discoverer of that little 'cave' in the wall of a river gulch to realise that he was in fact poking his head into a Miocene rhino, through its rump. A cast of the impression has been made and shows the appearance of the carcass. It is much bloated, which shows that the rhino had been killed some time before its entombing, probably by the gases or heat of the eruption. In the neighbourhood there are smaller cavities, which are impressions of tree stumps.

Great volcanic outbreaks may be traced in many other areas as well, for instance in southwestern Colorado where the San Juan Mountain massif is of volcanic origin, and in the Great Basin, the Colorado Plateau, and Mexico. Uplift of the Rocky Mountain area was resumed in Miocene times and was immediately accompanied by erosion; this process continued into the Pleistocene and led to the reappearance of the mountains that had been hidden under the Oligocene peneplane. The mammal-bearing early Tertiary deposits were preserved along the rims of the basins.

The Colorado Plateau was also uplifted in the early Miocene and was similarly attacked by erosion. The development of the Grand Canyon begins here, with the formation of a very respectable gorge during the Miocene. In the late Miocene the plateau was tilted to the northeast and established a new drainage pattern with internal basins receiving sediments.

The main mammal-bearing deposits of the Miocene in the United States are found in the Great Plains area. These formations are generally of coarser texture than the underlying Oligocene ones, reflecting the dramatic rejuvenation of the landscape, and hold great amounts of volcanic products. The early Miocene or Arikareean is represented by the Harrison beds of Nebraska with their remarkable *Daemonelix* or 'devil's corkscrews', giant spirals of harder rock eroded out of the softer, surrounding matrix. They are probably the filled-in burrows of some small mammal, perhaps a rodent (bones have in fact been found in the 'corkscrews'). Here belongs also the remarkable bone-bed of Agate Spring, Nebraska, where remains of *Diceratherium* and other animals are found by the thousand, all thrown together in utter confusion. The middle Miocene or Hemingfordian is represented by the Rosebud of South Dakota; late Hemingfordian fossils, with an especially rich horse record, are found in the Sheep Creek, Nebraska; while the Snake Creek of the same state dates from the late Miocene or Barstovian (it also contains Pliocene deposits).

Rich Miocene faunas are also known in other areas. The John Day formation of Oregon, consisting mainly of volcanic tuffs and overlain by the great basaltic flows of the Columbia River lava, dates from the Arikareean in its upper, mammaliferous part. Localities in Colorado (Pawnee Creek) and California (Temblor) have yielded Hemingfordian mammals, while Barstovian faunas are known from many localities including the Mascall of Oregon and the type locality at Barstow, California. An important marine fauna is known from Maryland. Florida, which had been generally inundated until the Oligocene, then emerged partially in the form of smaller or larger islands off the coast in the northern part of the present-day peninsula. One large island in early Miocene times became populated by a rich mammalian fauna, the record of which is preserved at the locality Thomas Farm.

The Miocene floras of North America are concentrated in the west, where they are numerous. Of special interest are those associated with mammalian fossils, like the Mascall flora from the early Barstovian, dated at 15·4 million years. Analysis indicates that this represents a climate not unlike that in the lowlands of the Mississippi basin and Atlantic coastal plain today, with some 'Appalachian' elements added. The area apparently consisted of wooded valleys with streams, swamps and small lakes, between hills with an open vegetation. Both plains and forest forms are found among the mammals.

In Europe, although the main Oligocene inundation had passed, inland seas still covered great areas in the Miocene. The Lattorf Sea in the north had retreated and Central Europe was now united to the Scandinavian-Russian land mass; there remained only a wide freshwater basin in eastern Germany and Poland, drained by a proto-Elbe River. Further west the Rhine Graben experienced new marine incursions and intense volcanism. Britain and the Low Countries were joined and formed a plain, while still further west Brittany was separated from the continent by a marine strait.

The Alpine molasse trench was still in existence and mostly covered by freshwater lakes, which were receiving abundant molasse sediments. The main arc of the Alps, the Apennines, Dinarids and Carpathians were already established as mountain ranges. Thus they closed off the wide expanse of brackish-water inland sea that covered the Pannonic basin within the Carpathians, southern Russia with the Pontic (Black Sea) basin, and extended east to the Aralo-Caspian basin.

The vicinity of the sea in most areas produced a humid climate, for instance in Ukraine where we find a deciduous-coniferous forest vegetation that now lacked the element of subtropical evergreens that had thrived here in the Oligocene. Even in the early Miocene there would be aridity in the lee of mountains, for instance in Rheinhessen and the north Carpathian foreland. The middle Miocene was uniformly humid. In the late Miocene, somewhat drier conditions gradually set in, and drought affected for instance the Vienna basin and Bessarabia. The gradual lowering of the temperature is evident in the composition of the floras, for example palms vanished from Europe north of the Alps during the Miocene.

The Mediterranean, which was now emerging as a remnant of the much greater Tethys, communicated with the Atlantic by way of the Rif Strait in

Morocco, north of the High Atlas that had been formed in Eocene times. In southern Spain the Betic Strait furnished another channel between the Atlantic and Mediterranean, while the area in between formed an island. But in the early Miocene a new sequence of mountain-building was initiated, leading to the complete emergence of Algeria and Tunisia. The Rif Strait and Betic Strait became folded and uplifted in the later Miocene, so that a land bridge between Africa and Spain was formed.

The Miocene succession in Europe begins with the Burdigalian age, represented by richly fossiliferous localities in the Vallés-Penedés basins near Barcelona, Spain. At this time, isolation from Africa in the south by the two straits and from the European continent in the north by the Pyrenees led to a stage of pronounced local differentiation, or endemism, in Spain. On the European continent, a great number of molasse and other open-air sites have yielded Burdigalian mammals, and there are also important cave and fissure fillings, for instance Wintershof-West in Bavaria with an extraordinary number of remains of small mammals. Next comes the middle Miocene or Helvetian which is also represented in the molasse trench and at such important localities as Sansan in Gers, France; but in the Vallés-Penedés this is the time of a marine transgression.

Another famous French locality, La Grive St Alban (Isère), gives us the fauna of the late Miocene, or Tortonian; its equivalent further east is termed the Sarmatian. The Helvetian and Tortonian are frequently united into a single age, the Vindobonian. In Spain, the Miocene record of the Vallés-Penedés is continued after the marine incursion of the Helvetian, by Tortonian or late Vindobonian faunas that now exhibit a prominent African and European element. Evidently this is due to the opening of a land bridge to Africa, and it may also be suspected that the Pyrenees now formed a less efficient barrier than in the Burdigalian. At the same time the climatic conditions in the Vallés show increased humidity with marshy forests replacing the gallery woods of the Burdigalian age, while the neighbouring Penedés basin had a drier aspect and shows evidence of torrential seasonal streams.

The Vindobonian molasse flora of Oeningen at Lake Constance is one of the latest with palms present north of the Alps: there are also crocodiles and other reptiles, birds, Vindobonian mammals, and various invertebrates; a total of some 1,500 species of plants and animals have been discovered. A celebrated find from Oeningen is the skeleton of a giant salamander, discovered in the eighteenth century, and originally regarded as the skeleton of a man drowned in the biblical Flood. These giant salamanders, *Andrias*, range from the Oligocene to the Pliocene in Europe.

They are also found in the Tertiary of North America, and the genus survives in the living giant salamander of East Asia, which may reach a body length of nearly 1·6 metres (over five feet).

Tectonic events in Europe in the Miocene were merely echoes of the tremendous earth storm that shook the Asiatic part of the Tethyan system and resulted in the folding and uplift of the main part of the geosyncline. In India the most intense phase of the Himalayan orogeny comes in the middle Miocene. Between the rising mountain chain and the Indian shield, now on the last lap of its long journey from ancient Gondwanaland near the South Pole, there developed a trough in which the accumulation of the Siwalik deposits took place. It has been suggested that these formations, which cover the entire sequence from the middle Miocene to the middle Pleistocene, were deposited by a single great stream – the Siwalik River – which may have originated in the Himalayas in Assam. From there it flowed to the northwest along the trough, then turned south and became the proto-Indus debouching in the Arabian Sea. A later reversal of drainage led to the expulsion of the Indus and the establishing of the Ganges in connection with the uplift of the Siwalik Mountains.

The Siwalik Series gives an important record of mammalian life in India, beginning with the Miocene Kamlial and Chinji formation; the later zones belong to the Pliocene and Pleistocene. Further west, early Miocene faunas have been found in the Bugti beds of Baluchistan. There are only a few additional localities for mammals in Asia, mainly in Mongolia, where the Miocene history of this area is closed with the rich Tung Gur horizon.

In the East Indies the greatest disturbance occurred in the late Miocene along the geosyncline curve of Timor, Celebes, and the New Guinea spine. The same is true for the Philippines and Japan, where the major folding and uplift also dates from the late Miocene. There is an interesting climatic succession in Japan. Early Miocene marine fauna here is of a cool-temperate type which may suggest a great influence of the cold Oyashio Current. This effect appears also to be reflected in the early Miocene floras. On the other hand, the transgressive sea of the middle Miocene in Japan indicates subtropical or tropical conditions with molluscs of mangrove type and algally secreted limestones containing corals and 'warm' foraminifera. Floras of the same date bear witness to a humid subtropical climate with elements of the so-called *Liquidambar-Comptoniphyllum* flora that now characterises the lowlands of Formosa some ten degrees further south. Probably the rise in temperature is due to a change in the oceanic circulation pattern, bringing the warm Kuroshio Current to bear on to the coast of Japan. Again in the late Miocene cooler conditions returned, perhaps due to the orogenic disturbance. In West Japan the fraction of subtropical elements in the flora is gradually reduced from one-third to one-sixth, while at the same time the subtropical marine fauna retreats southward and the cool-temperate forms advance.

Enclosed within the system of Tethyan geosynclines, the wide area of northern Asia formed a stable block, from which the regressions were gradually withdrawing the last remnants of the interior seas. Miocene uplift caused a disruption of the old sedimentation patterns which have given us such a rich record of the earlier Tertiary life of Mongolia, and with the Tung Gur comes the last major contribution of this area to the history of the Asiatic fauna.

In Africa the Miocene record is also scattered and incomplete. The most important faunas come from the early Miocene of Rusinga and Songhor in Kenya and Napak in Uganda; a late Miocene fauna from Fort Ternan, Kenya, is also notable. Marine faunas are known in some areas; a lower Miocene assemblage from Zululand contains tropical forms like the giant shark *Carcharodon,* the tropical molluscs *Strombus* and *Tonna,* the foraminifer *Amphistegina,* and the coral *Flabellum.*

The later Tertiary mammalian fauna of the World Continent was perhaps the richest that has ever existed on the face of the earth. It is as if mammalian life had been proliferating in ever increasing numbers, exploring every ecological nook and cranny that could be populated, testing how large or how small you could get, how best to adorn yourself with tusks or horns; how to fly, swim, climb, run, dig, jump, hunt, eat, kill and defend yourself, better than ever before. All the manifestations have a single keyword: adaptation. Most of the evolutionary lines of the later Tertiary had a fairly long history behind them: they had got far enough to attain basic adaptation for a given way of life. What remained now was to perfect it. And so, in the late Tertiary, the mammals were increasing in efficiency, under the constant supervision of natural selection – a perfectionist potentate. And in general this would also involve an increase in beauty, in gracefulness, in elegance.

Even in this splendid company, perhaps the most impressive creatures were the proboscideans. They emerged from Africa in the beginning of the Miocene to spread very rapidly throughout Eurasia, where they are found ubiquitously in the Burdigalian. Later on, in the late Miocene, the first proboscideans entered North America and rapidly proliferated there as well. Finally, in the Pleistocene, they colonised South America.

At the moment we are only concerned with the Miocene part of their story. The main line of the proboscideans is formed by the family Gomphotheriidae or pig-toothed mastodonts, most of which were long-jawed. Here belongs also the early *Palaeomastodon* from the Fayum. The mastodonts of this family are mainly characterised by the rounded cusps on their cheek teeth. It was the shape of these cusps that suggested to Cuvier, always a Frenchman, the term mastodont (tooth resembling the female breast). In general, the mastodonts were longer of body and head than

Figure 29. Apparently the commonest proboscidean of the Old World savannas in Miocene times, the long-jawed mastodont *Gomphotherium* also invaded North America. Shoulder height over 2·5 metres (8–9 feet).

true elephants, with shorter limbs; there were usually tusks in both the upper and lower jaws, as in *Palaeomastodon*.

There is however much variation on this general theme. *Gomphotherium*, which ranges from the Miocene into the Pleistocene both in the Old World and the New, resembled *Palaeomastodon* but was larger and had much longer jaws. The upper tusks were of medium length, while the lower ones at the end of the very long jawbone were quite small and chisel-like. Another Miocene member of the family is *Rhynchotherium*, another four-tusker in which the lower tusks are however relatively long and pointed; it has been regarded as a forest form, whereas *Gomphotherium* was mainly an inhabitant of the savanna.

The true mastodonts (family Mastodontidae) comprise the single genus *Mastodon* which dates back to the lower Miocene and also migrated far and wide, reaching North America in the late Miocene. Its best-known members are found in the Pleistocene. Also in Africa there appears in the early Miocene still another group of mastodont-like forms which may be ancestral to the true elephants; these are the stegolophodonts. In these the jaws tended to become shorter and the number of cusps in the teeth to increase. The stegolophodonts invaded Asia in the late Miocene.

A remarkably specialised line, which has a forerunner in the Eocene *Barytherium*, is *Deinotherium*, sole known member of the family Deinotheriidae. In these forms, some of which grew to enormous size, there were no tusks in the upper jaw, but the lower tusks were powerful and turned sharply downward forming a hoe-like apparatus. The deinotheres appeared in the Burdigalian in Europe and Africa, and reached Asia somewhat later, but never made it across the Bering bridge.

Another group of conspicuous forms, the rhinoceroses, continued to flourish in the Miocene, especially in Eurasia. To be sure the transverse-horned *Diceratherium* had become extinct in the Old World but, as we

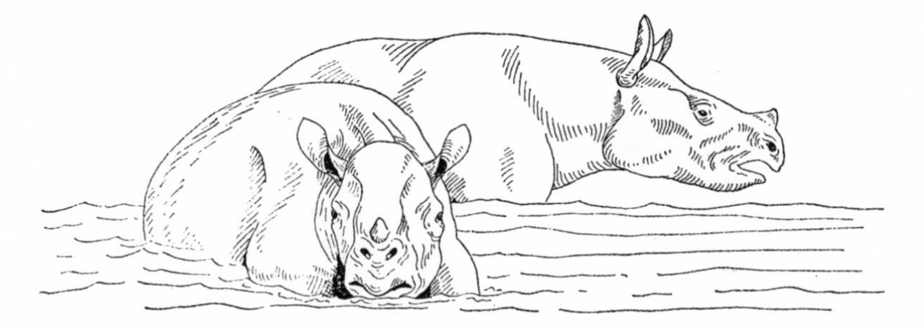

Figure 30. *Teleoceras* of North America is a representative of the hippo-like, amphibious rhinos of the Miocene and Pliocene, the ecological successors of the amynodont family. These rhinos were common in both hemispheres, *Chilotherium* being their chief representative in the Old World.

have seen, was still common enough in the New to leave its impression in a lava flow; it persisted into the middle Miocene. In late Miocene times the teleocerine rhinos made their appearance in North America: they were extremely long-bodied and short-legged forms, obviously amphibious in habits. In the Old World, various forms of Oligocene origin continued to flourish in the Miocene, for instance the hornless *Aceratherium* and the extant genus *Dicerorhinus*. The gigantic indricotheres survived in the early Miocene, then became extinct; but meanwhile another line of giant rhinos was arising in Spain, to culminate much later in the elasmotheres of the Ice Age.

Among North American perissodactyls the rhinos were over-shadowed by the horse family, for in the Miocene the equids entered upon a great phase of evolution and diversification. Although most of the Miocene horses remained browsing woodland forms with low-crowned teeth,

Figure 31. *Anchitherium*, the typical forest horse of the Old World Miocene, was originally an immigrant from North America. Shoulder height less than three feet.

continuing the *Miohippus* tradition, a few forms in the middle and late Miocene invaded the grassy plains that were spreading in North America and became adapted to grazing habits. This line starts with the genus *Merychippus*, in which we can observe a very rapid increase in the crown height of the cheek teeth, changing them into the veritable millstones for tough grass that are typical of modern horses. By the late Miocene, *Merychippus* was ready to radiate into several forms of plains horses, large and small, but all of them still equipped with the three-toed feet inherited from the woodland horses.

The plains horses did not succeed in crossing the Bering bridge in the Miocene, but the forest horse *Anchitherium* made the journey shortly after its appearance in North America and is found in the Burdigalian of Europe. It then persisted throughout the Miocene in both hemispheres.

Apart from rhinos and horses, perissodactyls were now rare in both hemispheres. There are a few tapirs and, rather more frequent, the big, slightly horse-like, but claw-bearing chalicotheres.

In fact at this time the artiodactyls were already superior in numbers to perissodactyls. The primitive families were gone or on the wane. In North America the gigantic *Dinohyus*, last, largest and most monstrous of all the entelodonts, is still found in the Arikareean. In Asia the anthracotheres still flourished and reached their peak of diversity in the early Miocene, but in Europe they were all gone in the later Vindobonian, and in Africa too their numbers dwindled nearly to extinction at the end of the Miocene. Instead the two modern groups, the swine in Europe and peccaries in North America, were taking over. The typical European and African pig of the time is the small *Listriodon,* in which the lower tusks are more developed than the upper, contrary to the condition in most pigs. Towards the end of the Miocene a large form of this genus entered eastern Asia, but generally speaking other types of pigs predominated in Asia. In

Figure 32. The chalicothere *Moropus* from the Miocene of North America and Asia, a large, claw-bearing perissodactyl. From a restoration by B. Newman.

the Caucasus there lived a remarkable giant hog with a unicorn-like bony horn on its forehead. The modern genus *Sus* appeared in the late Miocene.

It was in the Miocene that the Old World swine tended to develop the protruding canine tusks typical of that family. The New World peccaries or Tayassuidae may also have large canine teeth but they do not protrude sideways. This family never became so abundant or diverse as their Old World cousins.

By and large, however, the ruminants were now overshadowing the pig-like forms. In North America the oreodonts reached their acme in the early Miocene. The majority of the forms belonged to the family Merycoidodontidae, with something like sixteen genera in the Arikareean; after that the number was rapidly reduced and only two or three genera survived into the Pliocene. Some of the oreodonts were apparently amphibious, with a hippopotamus-like appearance. Others were hog-like, and in one group a small proboscis like that of a tapir was developed. Most oreodonts carried low-crowned cheek teeth, suggesting browsing habits, but in one form the tooth crowns became markedly higher and this may have been a grass-eater. There is also a remarkable group, very small but large-headed, which were clearly amphibious in habit; the shape of the skull shows that

Figure 33. Most grotesque of the North American protoceratid ruminants was *Synthetoceras* of the Miocene and Pliocene with its remarkable head adornment. These animals probably played the same ecological role as the giraffids in the Old World.

the animal could lie submerged with only its nostrils, eyes and ears above water, like a hippopotamus.

There was another, small family of oreodonts, the Agriochoeridae, which dates mainly from the Eocene and Oligocene although its last members are found in the Arikareean. These forms were rather slimly-built, with a somewhat elongated head; long in body and tail, they may have resembled carnivores a little. Probably all of the oreodonts were ultimately descended from ancestors of this sort of type.

The camels, close relatives of the oreodonts, were also limited to North America. They were now very varied and came in all sizes from the graceful little gazelle-camel *Stenomylus* to the big long-legged and long-necked giraffe-camels of the *Oxydactylus-Alticamelus* line. A remarkable find was that of a herd of *Stenomylus*: many of the skeletons were still in their death

Figure 34. Almost as peculiar as the contemporary *Synthetoceras* was *Cranioceras*, sporting a single occipital horn in addition to the two frontal horns; but in deer fashion the horns terminated in forked, deciduous antlers.

position, with limbs outstretched and head thrown back, while others had been pulled apart by carrion-eaters. It is thought that the animals died from lack of water in the dry season, the carcasses being later covered by the deposits of a subsequent flood, so becoming entombed in the Harrison Formation of Nebraska.

The higher ruminants or pecorans were represented by several families. The primitive North American hypertragulids, holdovers from Eocene and Oligocene times, were common in the Arikareean but then died out; most were of small size. A related but very spectacular American group is the Protoceratidae with, for instance, the bizarre four-horned *Syndyoceras* and the even more fantastically adorned *Synthetoceras* with its high Y-shaped nose horn in addition to the frontal pair.

A group of Eurasian and North American forms that may be ancestral to both the giraffes and the deer may perhaps be united in the family Palaeomerycidae. At first they were hornless, but in the later American forms bony horns tended to develop. The remarkable *Cranioceras* possessed two horns over the eyes and a third arising from the back of the skull. In some instances these bony horns may have been crowned by small antler-like structures and thus functioned in the manner of the pedicles of deer.

The antler is shed and rebuilt annually, while the pedicle is permanent. In this way perhaps the *Cranioceras* group may have given rise to the peculiar Miocene deer or semi-deer of North America, the dromomerycines, which had very long pedicles carrying small antlers at the end.

Still another series of horned ruminants in North America is formed by the prongbuck family, Antilocapridae, which appeared in the middle Miocene. The earliest antilocaprids were small animals probably not unlike the roe deer in appearance; although permanent, their horns were antler-like in appearance and forked or three-tined. But the main evolution of the pronghorn antelopes comes in the Pliocene and Pleistocene.

In Europe some primitive pecorans also survived in the Miocene, the most remarkable perhaps being the small *Cainotherium* which persisted into the Burdigalian of isolated Spain. The chevrotains or Tragulidae are represented in Eurasia by *Dorcatherium*, a small artiodactyl which in the flesh may well have looked like some large rodent, which is what the present-day *Tragulus* does until closer inspection reveals it to be a miniature ungulate.

The earliest deer in Eurasia belong to the genus *Dicrocerus*. They were small, with very simply built antlers usually forming a single fork; nevertheless they were successful and lasted from the Burdigalian to the end of the Pliocene. In the middle Miocene appeared forms in which the antler terminated in a small palmate disc.

The first giraffids occur in the Miocene but the group reached its main development in the Pliocene. A very odd creature of the Spanish Burdigalian is *Triceromeryx*, a three-horned animal resembling the American *Cranioceras* but probably not directly related to it; probably it was an offshoot of the giraffe line. The earliest bovids also appeared in the Burdigalian in the form of primitive woodland antelopes such as *Eotragus*. A few more were added later in the Miocene, among them the first true gazelles, genus *Gazella*.

The roster of land carnivore families was completed in the Miocene by the Hyaenidae. Burdigalian hyena-like viverrids are present in Europe and apparently were ancestral to the first primitive hyaenids of the *Palhyaena* group, found in the Vindobonian. Towards the end of the Miocene highly evolved hyenas (*Percrocuta*) are already found both in Europe, Africa and Asia.

The felid family continued to produce a range of forms from extreme sabre-tooths to 'normal' cats, the latter mostly being of moderate size, while some of the sabre-tooths became great, powerful monsters larger than a lion. Then there is the enigmatic *Hyainailouros*, which unfortunately is known only on the basis of a few outsize cheek teeth and jaw fragments (and which, it seems, was also the author of some gigantic fossil footprints found in the Burdigalian of Hungary). We do not really know if this immense form was a felid; it might even be a creodont. It has been found in Europe and Africa as well as Asia.

The mustelids flourished greatly in the Miocene. Martens are common and the modern genus *Martes* appeared in the early Miocene in Europe. The otter group is also common and varied, with many different forms apparently specialising in fish, crabs and molluscs. Early members of the polecat and weasel genus *Mustela* are found in the late Vindobonian. There are also badger-like mustelids, and some forms in Europe and North America are unmistakably skunk-like. Even the honey badger group of Africa is represented. The American giant form *Megalictis* in the Arikareean was the largest of all mustelids and reached the size of a black bear; it has been likened to a large primitive wolverine.

The dog family is even richer in species than the mustelids, especially in North America where it reached an all-time peak in the Arikareean. Many of the Miocene canids represent forms carrying over from the Oligocene without much change, for instance the large bear-dogs *Amphicyon* and *Dapheonus,* the small civet-like *Hesperocyon,* and the hyena-like *Osteoborus* which had a peculiarly inflated forehead or 'glabella' resembling that of the Pleistocene cave bear. The wolf series is represented in America by *Tomarctus,* a link in the chain leading from the Oligocene *Mesocyon* to the Pliocene *Canis.* In the Miocene of Europe there are dogs with a cat-like modification of the teeth. These were presumably highly predaceous animals, while the large amphicyons with their flattened molars may have been omnivorous.

Raccoon-like carnivores were present in both hemispheres, with *Sivanasua* as the European genus, while some five or six genera are found in North America. Among these may be noted the modern genus *Bassariscus* (ringtails) which appeared in the late Miocene. A panda-like form (*Schlossericyon*) is found in the Burdigalian of Spain.

The bears fall into two main groups. On the one hand there are the gigantic dog-bears (*Hemicyon*) which were still dog-like and probably digitigrade in spite of their great size. In the other group is the small *Ursavus,* which is much more bear-like in structure. Its history begins in the Burdigalian with forms no larger than a small dog. Late Miocene species were somewhat larger, but still smaller than the smallest of the modern bears, the Malay sun bear.

The smaller mammals also have a rich record in the Miocene. The opossum *Peratherium*, an Eocene survivor, made its last appearance in the early Miocene both in Europe and North America. Hedgehogs, moles and shrews were common in both hemispheres. Bats of many kinds are also found. Among rodents, primitive beavers were very common, and the same holds for the squirrels and the hamster-like forms. In Europe the dormice were at the peak of their history, with upwards of ten Miocene genera. In North America we find an endemic Mio-Pliocene family, the mylagaulids or horned gophers. They were fairly large, powerful digging animals with one or more bony horns on the forehead or nose, which may

Figure 35. Big dog-bears like *Hemicyon* are common in the Miocene of both hemispheres. They were more lightly built, and probably more carnivorous in habits, than most true bears.

have been used in defence or as a help in digging. A related family is the Aplodontidae with the living sewellels or 'mountain beavers' of North America. This family flourished in the Miocene, and (on the basis of a somewhat doubtful record) may have been present in Eurasia as well.

The last stragglers of the ancient *Paramys* group are found in the Arikareean, while the first members of the great rat and mouse family (Muridae) appeared in the Vindobonian of Europe.

In Europe are also found some scattered records of forms related to the scaly anteaters of Asia and Africa.

Adding up all the evidence, it seems that intermigration between Eurasia and North America occurred in the early, middle and late Miocene, but was at all times limited in scope, probably because the climate of the land bridge area was relatively cool. Nevertheless it led to a greater faunal similarity between the two hemispheres than in the Oligocene.

In Africa the opening of communications to Eurasia is clearly reflected in a great immigration wave comprising the hedgehogs, shrews, cats, dogs (with *Amphicyon*), chalicotheres, rhinoceroses, swine, palaeomerycids, giraffids, bovids and pikas right at the beginning of the Miocene. Perhaps the aardvarks should also be mentioned here; although they are now restricted to Africa, primitive forms are found in the Eocene of Europe and fully-fledged aardvarks in the Miocene.

The proboscideans of Africa comprise deinotheres, rhynchotheres, true mastodonts and stegolophodonts; the record is probably incomplete. Endemic African forms in the Miocene include the hyraxes, with dassies of modern type; most of the Oligocene forms had become extinct. There are also the golden 'moles' or Chrysochloridae; the elephant shrews, so-called not because of their size (they are very small hopping forms) but because they have a proboscis; and the tenrecs, giant tailless insectivores now restricted to Madagascar. The endemic rodent family Phiomyidae continued to flourish, and in addition we now find the related cane 'rat' *Thryonomys*.

In the early Miocene, Africa was still inhabited by numerous creodonts and there are even some fragments of a creature that might be a survival of the Arctocyonidae – the carnivorous condylarths that became extinct in the Eocene of the World Continent. An enigmatic group of creodonts from Kenya is that of *Teratodon*; these animals, the size of a fox or small dog, had very powerful crushing premolars that when worn down were somewhat like those of a hyena. But the majority of the creodonts in East Africa belong to the family Hyaenodontidae, of which no less than seven genera have been identified. Some of them, like *Hyaenodon* and *Dissopsalis*, are probably invaders, while some others are endemic African forms which evolved on the spot from much earlier invaders. In the later Miocene all these archaic forms were gone and there are true cats, dogs and civets all over the continent.

Then there are the primates. We left them in the Oligocene with the suggestion that at least six lines of higher primates were present in the Fayum; the number found in the Miocene is even greater. The roster of evolving forms is in fact very large and supports the idea of Africa as a homeland for the Old World monkeys and apes.

We even find prosimians related to the bush babies (*Galago*). The monkeys include the ancestral langur *Mesopithecus* and, in the late Miocene, a true macaque. There is an early gibbon-like form (*Pliopithecus*) and two larger ape species, both of which are now referred to the genus *Dryopithecus*. These apes are evidently ancestral to the chimpanzee and the gorilla respectively. An almost complete skull belonging to the proto-chimp 'Proconsul' has been found.

It may be remembered that the northern prosimians made their farewell bow in the Arikareean of North America (*Ekgmowechashala*). Shortly afterward, however, the new-fangled gibbons and other apes invaded Europe. Both *Pliopithecus* and *Dryopithecus* appear in the middle Miocene and persist into Pliocene times. Good skeletal material of *Pliopithecus* has been found in the Vindobonian of Neudorf (Czechoslovakia) and this has revealed many primitive characteristics. The body and legs were fairly long and slim and the arms only moderately elongated, and a small tail was present. Similarly, the Miocene *Dryopithecus* had comparatively short arms and long legs. The long arms of modern great apes and gibbons have

Figure 36. *Oreopithecus*, the mysterious man-like ape of the late Miocene swamp forests in Tuscany, Italy. With its long arms, it probably climbed hand over hand in the trees.

evolved as an adaptation to brachiation, or swinging hand over hand. The Miocene forms were probably brachiators that had not yet evolved extremely specialised limbs. The function had to precede the structure.

In the coalfields of Tuscany, Italy, a remarkable primate has been found, apparently dating from the late Miocene. This is *Oreopithecus*, perhaps a descendant of the tiny *Apidium* from the Fayum. This Tuscan ape, which is about the size of a chimpanzee, shows some curious man-like traits: the canine teeth are relatively small, the first lower premolar is also small and two-cusped, the spine and hip bone indicate an upright position, even the general shape of the skull and jaw looks somehow proto-human. It is fascinating to speculate on *Oreopithecus* as an 'experiment' in man-like adaptation, for despite all the similarities it seems quite clearly not related to man; its cheek teeth have a completely different cusp pattern. *Oreopithecus* may have been a brachiator. The arms are somewhat longer than the legs; at any rate it was arboreal in habits, judging by its environment, forested marshland. The small size of its canine teeth may suggest that it used artificial weapons, but there is no proof. After the end of the Miocene we have no further trace of this mysterious being. Perhaps one day we shall know more about its story and this would be of the greatest interest, for this is the only being that seems to show convergent evolution with man.

The human line itself seems to be represented in Miocene and Pliocene times by the genus *Ramapithecus* (the African material is often called *Kenyapithecus* but probably cannot be regarded as a separate genus). Possibly, remains of this group may be present in the Rusinga and Songhor faunas, dating from the early Miocene; in that case the temporal distance from the possible forerunner *Propliopithecus* is not great. As in *Propliopithecus*, the canine teeth were small and the anterior premolars man-like, but the teeth and jaws were even closer to the primitive true men of the earliest Pleistocene. In size *Ramapithecus* was something like a child of five or six, but unfortunately we only know the teeth, jaws, and a small part of the face. These suffice to show that the face was relatively vertical with a short upper lip, not with big protruding jaws as in the apes.

On the other hand we know too little of the postcranial skeleton to be able to say anything of the posture of *Ramapithecus* and its tool-using ability. That it did in fact use tools and weapons of a kind is of course highly probable; so does the living chimpanzee, and *Ramapithecus* with its feebler natural weapons would have had greater reason. On this point, circumstantial evidence may be forthcoming. Already there have been reports of fossil long bones from the African Miocene smashed as if somebody had been at them with a stone to get the marrow out.

In the late Miocene *Ramapithecus* is found at Fort Ternan, with a radiometric date of fourteen million years, and in Pliocene times it is also found in Asia and Europe. But here we must leave the proto-hominids for the moment to take a glance at the Miocene life in the seas.

Marine deposits with fossil mammals are known in many places. An important series of mammal-bearing strata was formed in the large bay that extended from the Atlantic into the French province of Aquitaine, and another series by the embayment of the North Sea that covered part of the Low Countries. The Calvert Cliffs of Maryland also contain an Atlantic fauna and in the south there are the marine beds in Patagonia. The coastal deposits in California and Japan have yielded an important record of life in the Pacific Ocean. Again the brackish-water sea of southeastern Europe is well represented. Indeed, as far as aquatic mammals are concerned, the Miocene record is very rich in contrast with that of the Oligocene.

Whales and dolphins (order Cetacea) are very common in Miocene marine deposits. Modern whales are characterised osteologically by a peculiar 'telescoping' of the bones of the skull, concomitant with the shifting of the nasal opening to the top of the head. This has occurred independently in the two great cetacean suborders, the toothed whales (Odontoceti) and whalebone whales (Mysticeti), both of which may be derived separately from the archaeocetes in which no telescoping had occurred and the blowhole still lay close to the tip of the nose. A few archaeocetes (of the family Dorudontidae) survived in the early Miocene but they are completely overshadowed in numbers by the odontocetes and mysticetes with a fully telescoped skull structure.

The main toothed group of the early Miocene was the family Squalodontidae, which evidently arose in the middle Oligocene; it dwindled rapidly in the middle Miocene and became extinct in the middle Pliocene. The squalodonts probably resembled modern dolphins in a general way but their teeth were unlike the simple peg-like structures of most modern odontocetes. The teeth in the front of the jaws were long and pointed, the back teeth on the other hand were serrated and blade-like and reminiscent of shark teeth. Many squalodonts had a long, pointed beak. There are also some related but even longer-snouted forms in the Miocene, such as *Eurhinodelphis* and its allies.

The modern family Delphinidae (true dolphins and ocean porpoises) arose in the early Miocene and soon completely overshadowed the squalodonts, with a fantastic climax in the middle Miocene – twenty genera or more. Another modern family which arose in the Miocene is that of the long-beaked river porpoises now living in the Amazon, Yangtsekiang and Ganges Rivers; the Miocene forms, however, were marine. This family is probably quite closely related to the squalodonts.

Another predominantly Miocene group which survives today is the beaked whale family, Ziphiidae; from a modest beginning in the early Miocene it evolved to a climax in numbers and variety in late Miocene times. In the ziphiids there is a tendency to lose all the teeth except for one or two tusks in the lower jaw. In the late Miocene there were also a couple of forms that may be referred to the small true porpoise family (Phocaenidae).

The largest toothed whales of the present day, the sperm whales or cachalots, also date back to the Miocene. Throughout the epoch, numerous members of this family are recorded. Many of them were still relatively small and evidently close to the original delphinid stock, but there are also a few forms approaching the true sperm whale in structure if not in size; these include species of the modern genus *Physeter*.

Among whalebone whales, the extinct family Cetotheriidae predominates. They arose in the Oligocene and there are about a score of genera in the Miocene. Very few survived in Pliocene times and the last died out well before the end of that epoch. Most cetotheres were much smaller than the main present-day forms but a few attained lengths of up to thirty feet; all had relatively short baleen. They have been likened to the present-day California grey whale, which is evidently descended from them. It has been thought that the cetotheres were favourite prey for the immense white shark *Carcharodon* which reached about the same size.

In the late Miocene some modern types of mysticetes begin to show up, all of them belonging to the rorqual family, Balaenopteridae; among them may be noted humpback whales as well as the modern genus *Balaenoptera*. As regards the right whale family, Balaenidae, there is one record of an early form in the early Miocene marine beds of Patagonia; otherwise right whales are unknown prior to the Pliocene.

The second large group of water-living mammals, the seals or Pinnipedia, enters the fossil record in the Miocene. This is true for all of the three families recognised: the Otariidae or sea lions, the Obodenidae or walruses, and the Phocidae or true seals. It is now thought that the last-mentioned evolved from otter-like ancestors, while the sea lions and walruses show more resemblance to the bears and may have arisen from some one of the primitive bear-like dog types that were so common in the Oligocene. As regards the otter-like ancestry of the Phocidae there is some corroborative evidence, for instance the presence of a very peculiar marine otter with seal-like characters which has been found in the early Miocene of California. The European otter *Potamotherium*, so common in the Oligocene and early Miocene, also had many seal-like characters. Although these forms in themselves may not have been ancestral to phocids, they indicate that such a trend was not unusual. The trend towards a marine life is, of course, illustrated by the sea otter and the polar bear today.

The geographic distribution of these forms is interesting. In the Miocene the sea lions appear to have made the northern Pacific their home, while walruses are found only in the Atlantic. Finally, the Phocidae appear to have evolved in Europe.

Among the Miocene sea lions are found types not unlike their present-day relatives, but the very big *Allodesmus*, although a sea lion by descent, evolved in a very aberrant fashion and came to resemble the elephant seal, which is a phocid. It had a proboscis, it reached a very large size, and apparently it was covered with blubber and lacked fur. This form is middle Miocene. In the late Miocene of California, a great change occurred; the early otariids are gone and we find modern types of sea lions and fur seals. It has been suggested that this is due to the progressive cooling of the water, which would bring in cold-adapted modern forms from the north. (See Plate 5.)

The Phocidae are fairly common in Europe in the later Miocene, both in the interior sea of the southeast and in the Belgian and Aquitanian embayments. They include ancestors of the ringed and Caspian seals as well as the monk seals and extinct forms.

Among the subungulates we have already met one aquatic order, the Sirenia, which were also common in coastal waters in Miocene times. In addition, however, we now find another order, the Desmostylia, which also appears to belong to the subungulates. Their known history extends approximately from the latest Oligocene to the middle Pliocene and they were distributed along the shores of the North Pacific Ocean; skeletons have been found both in Japan and western North America. The desmostylids were quite large and heavy animals which, unlike the sea-cows, possessed well developed limbs and probably moved on land in about the same manner as the sea lions. The jaws were more or less elongated with four tusks; the arrangement and the whole skull shape in *Desmostylus*, for instance, is not unlike the mastodont *Gomphotherium*. The type of tooth replacement also resembles that in sirenians and proboscideans. If of subungulate origin, the geographic and temporal separation of the desmostylids from the homeland of the subungulates, Africa, seems puzzling. However, as already mentioned there is now some evidence of an early

Tertiary migration of primitive subungulates to Eastern Asia. Alternatively, they may have migrated by way of the Bolívar Trench.

Many oceanic birds are known in the Miocene. Special mention should be made of the penguins. The earliest penguins are found in the Eocene; the group then built up through the Oligocene to a climax in Miocene times, with a size range up to giant forms as tall as a grown man. Also, the saw-toothed birds, which we have already met in the Eocene, were now very common, especially on the Pacific coast of North America; the large *Osteodontornis* had a wingspread of about five feet. The 'teeth', as may be remembered, were only spiky outgrowths of the jawbone.

Table 8. The Miocene of the World Continent

Epoch	Africa (faunas)	Asia (faunas)	North America (ages)	Europe (ages)	Date (Million years)
Pliocene	Bled Douarah		Clarendo-nian	Vallesian	
					10
Miocene		Tung / Chinji		Tortonian	
					12
	Fort Ternan	Gur Kamlial	Barstovian	Helvetian	
					16
			Heming-fordian		
	Rusinga	Bugti	Arikareean	Burdigalian	21
				Aquitanian	25
Oligocene	Fayum	Hsanda Gol	Whitneyan		

Table 9. Number of mammalian families in the Miocene of the World Continent

	Late Oligocene	Miocene		
		Early	Middle	Late
Total number of families	77	95	88	92
First appearances	5	26	4	5
Last appearances	8	11	1	1
Percentage newcomers	6½	27½	4½	5½
Percentage extinctions	10½	11½	1	1

It remains to take stock of the Miocene, a long epoch dominated by earth unrest, the uplift of sky-high mountains, and the vanishing of ancient Tethys, which also gives the first warning of future glaciations. It brought all parts of the World Continent, including Africa, into closer communication and exchange of land animals. It saw a succession of rich mammalian faunas with an essentially modern composition on the family level; the number of families does not fluctuate much and except for some early extinction there is very little turnover. The main lines of adaptation have been determined and we may see the beginning of a process of perfectioning. The continued evolutionary trends in the Miocene and Pliocene have often been interpreted as resulting from some inner urge towards a given goal, so-called orthogenesis. In actual fact it is due to long-range constant selection, or orthoselection. It has been claimed that evolution moves in a straight line but this is an optical delusion; close scrutiny of evolving lines always reveals deviations and fluctuations. These arise because the organism is always adapting to its current environment and mode of life, not those of the distant future. That the long-range trends are nonetheless real is because the populations tended to keep approximately the same ways of life for a long time and thus were constantly under the same general type of selective fashioning.

Partaking in this process we now find early representatives of our own zoological family, the Hominidae. This is perhaps one of the most important facts realised in recent years.

1 A valley in the Rocky Mountains 58 million years ago, in late Paleocene times. The basins were close to sea level and clothed in luxuriant deciduous forests. Heavy-tailed pantodonts of the genus *Barylambda*, about the size of a cow, in foreground; in middle distance, *Champsosaurus*, a long-nosed, gavial-like reptile, slipping into the water.

2 The Geiseltal brown-coal basin 48 million years ago, in the middle Eocene. Left, part of gallery forest with redwood and palms, encircling basin; willow and swamp cypress nearer lake. Two big lophiodonts, tapir-like primitive perissodactyls, are part of herd in distance; one is browsing on myrtle, the other is at a pond where it is stalked by panzer crocodile. The dry season is under way and the evaporating pond is fringed by rotting fish. Animals in right foreground are a small species of *Palaeotherium*, somewhat horse-like creatures about 3 feet in length. Agama lizard in left foreground; condors in sky.

3 Banks of proto-Nile at the Fayum, 35 million years ago, at beginning of Oligocene. Big four-horned animals wallowing in mud and standing in water are *Arsinoitherium*; between them, three giant dassies (*Megalohyrax*) are coming down to drink. Herd of anthracotheres in distance.

4 Oligocene plain in North America 32 million years ago. Dry riverbed is crossed by herd of primitive camels (*Poebrotherium*); there are also a running rhino (*Hyracodon*, distance left) and oreodonts of the genus *Merycoidodon*. One of the latter has been killed by a couple of sabre-toothed cats (*Hoplophoneus*). Small dog of *Cynodictis* type may have played the same role as today's jackals.

5 Skerries in the North Pacific in the middle Miocene, 20 million years ago. The plant-eating amphibious mammal *Desmostylus* is retreating from territory of big *Allodesmus* male. Related to the sea-lions, *Allodesmus* evolved to resemble present-day sea-elephants, although there is no direct relationship between the two.

6 A European forest 10 million years ago, in early Pliocene times. In foreground primitive deer (*Dicrocerus*); further off okapi-like giraffids of the *Palaeotragus-Samotherium* group. Otter is *Limnonyx*.

7 Peculiar, one-horned bovid, *Tsaidamotherium*, was a member of *Hipparion* fauna in Asia, 8 million years ago. In distance antelopes and *Hipparion* horses. The rich plains fauna of the earlier Pliocene marks the climax of mammalian history in the Old World.

8 A scene in Australia about 2 million years ago, with giant wombat (*Phascolonus*) and short-tailed kangaroos (*Sthenurus*). Trees are southern beech (*Nothofagus*) and monkey-puzzle (*Araucaria*). Cormorant in foreground.

9 Twenty million years ago in Patagonia. Santa Cruz Miocene with *Astrapotherium* (large animals on bank and in river, also herd on far bank) and the horse-like litoptern *Diadiaphorus*, pursued by marsupial carnivore *Borhyaena*. Heavy volcanic activity characterized the Miocene in many parts of the world.

10 Tar pit at Rancho La Brea, California, 15,000 years ago. Pack of dire wolves are trying to drag mired bison out of tar; one has fallen in. Big late-Pleistocene coyote turns to glance at approaching shasta ground sloth (*Nothrotherium*). Mastodont herd beyond second asphalt pool. Skull in left foreground is sabre-tooth *Smilodon*.

11 Neandertalers butchering reindeer in ice-age Europe, 40,000 years ago. Fleeing, woolly rhinoceros (*Coelodonta*); mammoth herd in distance.

Chapter 6

The Pliocene: epoch of climax

With loss of Eden.
J. Milton: *Paradise lost.*

THE UNSTABLE Miocene epoch with its worldwide upheavals was followed by a long serene interval: the Pliocene. It is as if we were emerging onto a wide grassy plain under a strong sun, with a rising wind from distant snow-capped mountains. Such a landscape, populated by great herds of three-toed horses, mastodonts and antelopes, might indeed be taken as a symbol of the Pliocene, just as the Pleistocene might be summed up in an Ice Age cave panorama, or the Miocene by a scene from the marshy molasse trench of the rising Alps, with its dense forests.

A comparatively short epoch, the Pliocene lasted some seven–ten million years, depending on where you count and how you count. There is no agreement on when the Pliocene began, or when it ended. Students of North American land mammals generally place the Mio-Pliocene transition at the Barstovian-Clarendonian boundary, and since the oldest Clarendonian date is 11·7 million years and the youngest Barstovian (late Miocene) 12·3 million years, twelve million should be a good approximation. The Clarendonian is then followed by the Hemphillian or mid-Pliocene which lasted up to about five million years ago (5·2 is the youngest Hemphillian date) and finally by the Blancan which carries us up to 1·5 million years ago, when we are well in the Pleistocene by other ways of reckoning. This may make it necessary to draw the Plio-Pleistocene boundary somewhere within the Blancan. The resulting confusion may be overcome by dividing the Blancan into two ages, the Rexroadian or late Pliocene, and the Blancan proper or early Pleistocene, with the transition at about three million years. Not all authorities, however, will accept such a solution.

If on the other hand you are a student of European land mammals you may prefer to draw the Mio-Pliocene boundary at the base of the Pontian age, with the immigration of the three-toed horse *Hipparion*. Most students of intercontinental correlations now believe that the Pontian began considerably later than the Clarendonian, perhaps only about ten million years ago which is close to the end of the North American early Pliocene. But many European authors regard the Pontian as the late Miocene and place the epoch boundary at the transition between the Pontian and the succeeding Astian age. The exact date of this is very obscure but may be of the order of five million years. If you then let your Pleistocene begin with the Villafranchian, dated at three million years, you wind up with a miniature two-million-year Pliocene which does not seem to be much use to anybody.

Again, if you work on Mediterranean marine faunas, your Mio-Pliocene transition will probably fall somewhere within the Pontian as recognised by land mammals, and confusion will be supreme.

But we must leave these problems to the specialists, and proceed to the geography of the Pliocene. North America had outlines corresponding closely to those of the present day. In the southeast there were still marginal inundations, especially the coast of North Carolina and large parts of Florida; only the northern part of the peninsula was emergent. In California the San Joaquin Valley formed a large bay, protected seaward by the rising Coast Ranges. Later on in the Pliocene the seas retreated to the west.

The Appalachians in the east, and the Rockies in the west, were still being elevated and eroded. Continued upwarping of the Colorado Plateau led to a gradual deepening of the Grand Canyon. But on the whole, the Pliocene was a time of quiescence between the paroxysms of the Miocene and the Pleistocene. The movements of the Cascadian revolution had abated, and the Sierra Nevada remained a chain of modest hills up to the end of the Pliocene.

There is a very large number of Pliocene mammal-bearing deposits in North America; at some of the sites fossil floras and faunas are found together. Many important localities are in the Great Plains region. The type sites Clarendon and Hemphill are both in Texas. A very important series of localities is located in the west coast states, especially California and Oregon. Many have been dated radiometrically, giving a detailed calendar for the Pliocene as outlined above. The Rexroad type locality is situated in southwestern Kansas where it initiates an important sequence of early Pleistocene strata.

The Pliocene floras tell of increasing aridity. Grass and shrublands conquer wide areas, and in the later Pliocene the northern part of the Great Basin was probably receiving only about 400 mm. rainfall per annum which is only about one-third of the corresponding amount in the Miocene.

The southern part of the basin was still drier. A weather forecast for 'a summer's day at Middlegate' (early Pliocene, Nevada) reads: 'Continued clear with afternoon showers'. Present-day conditions with rainfall of between 100–200 mm. were probably reached in the late Pliocene or early Pleistocene, when the uplift of the mountain ranges was resumed.

In Europe there was also a tendency for the sea to recede and the land to emerge. A shallow brackish-water inland sea still extended from the Aral basin westward across the Caspian and Black Sea (Euxenic) basins. In the early Pontian the sea also covered the Dacic basin between the Transsylvanian Alps and the Balkan mountains, but in the course of the Pontian this basin was cut off to become a freshwater lake fed by the Danube. This great river also flowed through the Pannonic Basin enclosed

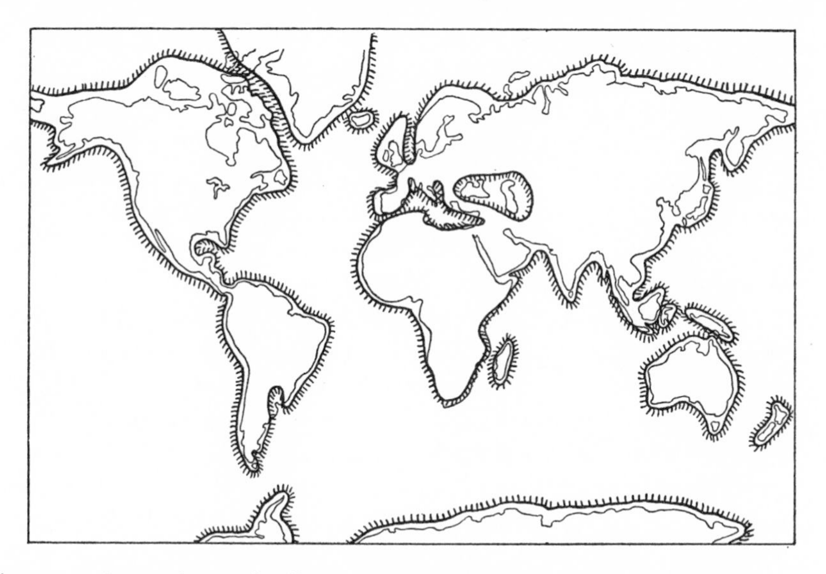

Figure 37. Approximate distribution of land and sea in the Pliocene epoch.

within the Carpathians and Transsylvanian Alps. In the later Pliocene the Alps and Carpathians were gradually elevated once more by simple uplifting without folding. As a result, the Danube, connecting the Pannonic and Dacic basins, was forced to cut a gradually deepening gorge which now forms the Iron Gate.

Of the ancient Oligocene inland sea there now remained only a freshwater basin in the German-Polish area, which was covered by lakes and peat-bogs; they have formed a rich brown-coal supply. The North Sea was a shrunken water body and the Channel was dry land; but the tip of Brittany was still separated from the continent by a strait.

Later in the Pliocene, the old Miocene peneplain of northwestern Europe became elevated with consequent renewal of erosion and dissection. A temporary communication between the North Sea and the Atlantic was formed by a channel in southern England, but at the end of the Pliocene it apparently silted up again.

In the Central Plateau area of France, and in the Siebengebirge, volcanism was going on throughout the Pliocene. The molasse trench at the front of the Jura Mountains formed a marshy area. The Aegean basín was also the site of a freshwater lake, south of which there was a continuous land bridge from Greece to Asia Minor. Some of our best Pontian faunas come from this region.

In Pontian times, the Straits of Gibraltar came into existence through downfaulting. Thus the land bridge from Africa to Spain was broken and Mediterranean and Atlantic water could blend once more. A large part of the northwestern Mediterranean was dry land in the Pontian; this land mass, called Tyrrhenis, comprised the present-day Balearic Islands, Corsica and Sardinia. Towards the end of the Pliocene it was cut up by grand-scale collapses which left these islands as relicts; perhaps this subsidence was connected with the rise of the live Mediterranean volcanoes of Vesuvius, Etna and the Lipari Islands and of the extinct volcanoes of the French Riviera.

The earliest Pliocene mammalian faunas in Europe are typical forest faunas. This early forest phase of the Pontian is sometimes referred to as a distinct age or sub-age, the Vallesian. Later on there was a great spread of grasslands and in the late Pontian or Pikermian sub-age the greater part of southern and eastern Europe was savanna or steppe.

A good example of the forest flora in Europe during the Pontian is given by the rich flora of Frankfurt on Main. Its remains have been unearthed from the Pliocene river bed of the proto-Main. The fossils come from a fairly wide area, probably the entire basin enclosed between the Taunus in the northwest, the Spessart hills to the east and the Odenwald in the south.

The river evidently was relatively sluggish, as it contained pond weeds (*Potamogeton*), water lilies and hornwort (*Ceratophyllum*). The shores were fringed by a rich reed, bulrush and sedge vegetation, while in some areas the swamp cypress and *Nyssa* would grow. The woods of the river valley consisted of poplar, ash, *Prunus*, hornbeam, beech, oak, elm, birch, tulip-tree (*Liriodendron*) and sweet-gum, with some additional maple, walnut and hickory species. The trees wore festoons of grape vines, ivy, snake gourd (*Trichosanthes*) and other lianas, while the undergrowth was formed by sweet-gale (*Myrica*), hazel, berberis and bladdernut (*Staphylea*). Higher up on the south-facing slopes of the hills, conifers would predominate, mixed with *Engelhardtia, Pterocaria,* magnolia and other southern and eastern forms. Higher up on the hills there would be a purely coniferous wood with spruce, small-coned pine, silver fir (*Abies*) and hemlock fir (*Tsuga*).

This flora is clearly akin to such Miocene floras as that of Oeningen, but many tropical forms are now lacking, for instance the palms. This is proof of a further reduction in temperature. Another interesting character of this flora is the number of modern species found in it, amounting to some seventeen per cent. In a late Pliocene flora from the Harz area this percentage rises to forty-four, and in the mid-Pleistocene Cromer flora to ninety-five. At the same time there was a gradual weeding out of the forms with mainly American or East Asian affinities, which had been completely predominant in the early Tertiary. At Frankfurt more than half of the species still belong to this group. In the late Pliocene the percentage had dwindled to twenty-seven, and in the Cromer beds to less than one per cent. The abundance of North American forms in the early Tertiary is presumably an inheritance from the days of Laurasia, the northern supercontinent; as an indigenous European flora evolved, this ancient element was gradually reduced.

The lowering of temperature is a continuation of a trend which started in the Eocene or even earlier. Now, however, there is clear evidence of a short-term oscillation superimposed on the general trend. For instance, a pollen diagram from the Pontian brown-coal deposits of Limburg in the Netherlands indicates a swing from cold to warm and then back to cold again. The pollen found in these deposits comes in part from the swamp forest, in part from the surrounding upland vegetation. The latter shows a remarkable change. In the earliest deposits, pine is predominant, but there is a gradual increase of deciduous broad-leaved trees – oak, beech, hornbeam. Next comes a zone where these trees reach their maximum, together with evergreens; a notable element is the mainly tropical sweet-leaf (*Symplocos*). Finally, cooler conditions return with an increase of *Tsuga*, pine and other conifers, while the deciduous trees dwindle and the hornbeam actually disappears. The whole sequence is startlingly similar to the floristic history of an interglacial during the Ice Age – only at this time, inland glaciation in Europe was still many million years in the future.

The earliest Pliocene faunas in Europe, although dominated by Miocene hold-overs of woodland type, have some diagnostic new forms, the most notable of which is the three-toed horse *Hipparion*. This early Pontian is especially well represented in Catalonia, Spain, where it is singled out as a separate age or sub-age, the Vallesian. Here belong also some faunas from Bessarabia and southern Russia, where the local Sarmatian may be a correlative of the Vallesian, and probably also the Oued el Hammam faunas in Algeria. The last-mentioned shows very clearly that the immigration of *Hipparion*, and thus the beginning of the land-mammal Pliocene, occurred long before the end of the marine Miocene: the continental mammal-bearing beds here are overlain by marine deposits with a typical Tortonian fauna. The immigration of *Hipparion* is recorded in a sequence of faunas in Bled Douarah near Gafsa, Tunisia.

Later Pontian faunas are especially numerous. This is the classical or typical *Hipparion* fauna, which Osborn in 1910 called 'the most famous, the most widely distributed, and the best known of all the mammalian faunas of the Old World'. Among the European localities may be noted Pikermi near Athens, from which has been derived the name Pikermian for the later Pontian; Samos, once part of the land bridge between Greece and Turkey; many sites in Hungary, the Vienna basin, southern Germany and France; the Vallés-Penedés basin in Catalonia; and a very rich sequence in the Teruel-Calatayud basin in Aragon. Other rich localities are found in Portugal, Italy, Turkey, and North Africa (Morocco, Tunisia and Libya). (Plate 6.)

The record of the late Pliocene, or Astian, is somewhat less impressive. Classical European sites are Roussillon and Montpellier in southern France. Other localities are known from Spain, Germany, Hungary, Rumania, etc. The fauna suggests a return of woodland habitat after the drier interval of the Pikermian.

In Africa only a few late Pliocene or Plio-Pleistocene mammal-bearing deposits are known. Here may be noted the Wadi Natroun in the Nile Valley, the diamond fields of Namaqualand in South West Africa, Omo in Ethiopia and the Kanapoi fauna by Lake Rudolf, Kenya.

Asia has a rich Pliocene record. The Pontian *Hipparion* fauna appears to be present nearly everywhere. Rich localities in Turkey and Persia bridge the distance between Europe and India, where the Siwalik Series contains a continuous record of Pliocene life. The sequence begins with the Miocene Kamlial beds, which are overlain by the Chinjis in which *Hipparion* may make its first appearance. The majority of the Chinji mammals are however of Miocene type, which has led to disagreement about the correct date of this zone. As we have already noted, the Vallesian or earliest Pontian faunas in Europe also have a distinctly Miocene stamp. Apparently, then, the Chinji may be partly equivalent to the Vallesian.

Above the Chinji follows the Nagri zone, probably a correlative of the middle Pontian (early Pikermian) in Europe, and then the Dhok Pathan zone which had yielded fossils of both Pontian and Astian affinities. This apparently takes us right up to the end of the Pliocene epoch, for the zones that follow are already Pleistocene in age.

East Asian Pliocene mammal-bearing deposits are especially plentiful. In the Chinese provinces Shansi, Shensi, Kansu and Honan, innumerable sites replete with fossil *Hipparion* fauna have been found; some of the richest are clustered in the neighbourhood of Paote, Shansi, and most appear to be equivalent to the Pikermian of Europe. The flora of Taiku in Shansi is typical of an arid climate, but in Honan a rich fauna with browsing herbivores suggests a forested biotope. There is a clear distinction between the plains fauna in the north and northwest, with a profusion of antelopes, and the southeastern woodland fauna with giraffids and browsing gazelles with low-crowned cheek teeth. *Hipparion* the three-toed horse was at home in both types of environment. (Plate 7.)

It seems clear that Asia was affected by the tendency to increasing aridity. Great areas in the interior of the continent were steppe or desert, gradually intergrading into savanna and finally into rain forest in the monsoon areas. But the climate was clearly warmer than now, for a rich temperate-type fauna and flora existed in present-day Siberia. In Japan a Pliocene flora indicates a very humid climate with foggy summers and mild winters.

Late Pliocene deposits are fewer, but there is a good record at some localities, for instance the Yushe basin in southeast Shansi.

The Pikermian may well be regarded as the climax of the entire Age of Mammals, at any rate in the Old World. It has been shown that the savanna environment is capable of supporting a greater biomass of mammals than any other habitat on land – much greater than the rain forest with its exuberant vegetation. For instance, the Tano Nimri Forest of Ghana supports less than 100 kg. mammalian biomass per square kilometre. The corresponding value for the Serengeti savanna is more than 4,500 kg., and an estimate for the Rwindi-Rutshuru plain in Albert National Park is over 23,000 kg. per square kilometre. Such conditions now obtain only in a few areas, mainly in Atrica, but evidently were very widespread in Europe, Asia and Africa during the later Pontian. These habitats then carried immense herds of antelopes and other bovids, giraffes, hipparions, rhinoceroses and mastodonts, while the old Miocene browsing herbivores were dying out, or became restricted to the shrinking wooded areas.

The horses are so important in Pliocene stratigraphy and evolution that it is only fair to begin the presentation of Pliocene mammals with this group. The Pliocene may indeed be regarded as the climax of the equid family. During the Miocene, the until then rather uniform horse group in North America had divided into two main branches. On one hand there were the browsing woodland horses in which the cheek teeth remained low-crowned, or brachydont. Two forms in this group managed to colonise the Old World, *Anchitherium* in the Burdigalian and *Hypohippus* in the late Miocene. On the other hand there evolved a group in which the cheek teeth tended to become high-crowned, or hypsodont. The evolution of these mainly plains-living, grazing horses begins with the genus *Merychippus* in the Miocene.

By Clarendonian times the hypsodont horses had become very diversi fied, while the brachydont horses were on the wane and became extinct in the Hemphillian. Some of the hypsodont horses, *Calippus* and *Nannippus,* became very small; these gracile miniature horses probably led a life not unlike that of gazelles. They retained three-toed feet, and this was also true for the larger hypsodont horses belonging to the *Hipparion-Neohipparion* group. The upper cheek teeth of *Hipparion* are usually easy to identify on the roundish or oval separate inner column or protocone, forming an isolated loop in the enamel pattern on the grinding surface. At some time in the Clarendonian *Hipparion* crossed the Bering bridge, then very rapidly colonised the Old World.

Before turning to the Old World hipparions, however, it must also be noted that there was another group of large, hyposodont horses in North America during the Pliocene: the *Pliohippus* group. This is the only line in which the side-toes tended to become further reduced and finally were lost. Exactly when in the Pliocene this transition occurred is not yet definitely known. In some populations there would be individuals having vestigial side toes living together with individuals in which the toes were absent and only the splint bones remained, as in modern horses. The teeth of *Pliohippus* are easy to distinguish from the hipparions because the protocone is generally attached to the other cusps. Such is also the case in modern horses, which are descendants of *Pliohippus.*

The *Hipparion* group never became very prominent in its homeland but enjoyed a tremendous success in the Old World, where it rapidly eclipsed and ousted its brachydont predecessors. In the earliest Pontian there appears to have been a rather uniform *Hipparion* in Asia, Europe and North Africa, probably still very like the original Bering migrant. These hipparions are fairly large, with a shoulder height of four feet or so, and with robust, stocky limbs; they are found both in woodland and plains associations. Later on in the Pontian, however, the hipparions gradually split into distinct grassland and woodland forms, specialising for various ways of life. Some tended to large size, others are small, some became very slender-limbed while others were heavily-built. In Spain a local dwarf type evolved, no larger than the American 'gazelle-horses'.

With the end of the Pontian, the hipparions tend to decline in variety; only a few, mostly large forms survived, and the number of individuals seems also to have diminished greatly. And so, even though hipparions were still in existence in the early Pleistocene, the true *Hipparion* fauna is limited to the Pontian.

Among perissodactyls, only the rhinoceroses continued to prosper in a

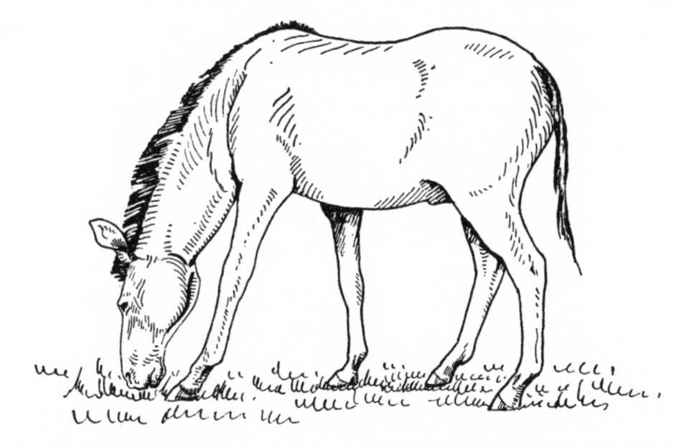

Figure 38. *Pliohippus*, first of the one-toed horses, arose in the Pliocene of North America. From ancestors of this type evolved the present-day horses, zebras and asses.

manner comparable to the horses and they were now mainly restricted to the Old World. The group was dying out in North America and only the amphibious, short-legged *Teleoceras* survived to the end of the Pliocene in this continent. But Eurasia remained rhinoceros land, and there are also several African records of this group. The European *Hipparion* fauna still included ancient *Aceratherium*, but in the late Pliocene it was gone and we find instead the first members of the modern African black rhinoceros group (*Diceros*). *Dicerorhinus* was still common both in Europe and Asia.

In East Asia the most common rhinoceros was *Chilotherium*, apparently a relative of *Teleoceras* and like it a short-legged, hornless type. We have also to note the gigantic *Sinotherium*, an elephant-sized animal with very highly developed, hypsodont cheek teeth somewhat resembling those of the horses. This evolutionary line may be traced to a form in the Miocene of Spain, and was to culminate in the Pleistocene *Elasmotherium*.

The two other perissodactyl families were already becoming relicts. The tapirs, though still present in North America, Asia and Europe, belonged to the single living genus *Tapirus*. The chalicotheres – those peculiar animals that might be thought of as giant clawed horses – became extinct in North America after the early Pliocene, but still formed a striking element of the Eurasian *Hipparion* fauna. The last chalicotheres in Europe died out at the end of the Pliocene; in Africa and Asia the group lasted into Pleistocene times. When the huge claw-bearing feet of *Ancylotherium* were first discovered at Pikermi, they were ascribed to some gigantic edentate like the ground sloths of South America.

The ascendancy of the Artiodactyla continued in the Pliocene, when the modern families were coming into their own. The deer multiplied in the forested areas of Eurasia; besides surviving Miocene types and others which became extinct, many new forms appeared related to the roe deer, the red deer, the muntjaks and the elks. But it was only towards the end of the Pliocene that true cervids appeared in North America to replace the aberrant antlered forms that had evolved there in the Miocene.

Mainly a southern woodland group, the Giraffidae also prospered in the Old World Pliocene. Most were 'normally' built and lacking the elongation of neck and legs seen in the living giraffe; the okapi is close to the central giraffid type. This type is exemplified in the Pontian of Eurasia and Africa by the widely distributed *Palaeotragus*, a moderate-sized, brachydont browser with two short, pointed bony horns. The closely related *Samotherium*, which was somewhat larger and more hypsodont, was presumably a grazer and inhabited the Old World steppes and savannas. Both forms ranged northward to Pavlodar in the Irtysch basin, Siberia, which testifies to the mild climate of the Pontian. The modern genus *Giraffa* is also present in the Eurasian *Hipparion* fauna; it is a mainly savanna-living treetop browser. This genus apparently did not reach Africa until the Pleistocene.

Other, more aberrant forms include *Giraffokeryx* with four long horns, and the very large *Helladotherium*, biggest of all the Pontian ruminants, a great hornless creature with somewhat camel-like proportions. Towards the end of the Pliocene the giraffids declined sharply in numbers, probably due to the increasingly cold climate.

All the other artiodactyls were however eclipsed by the fantastic proliferation of the Bovidae, which seem almost to explode into a spectrum of different forms in the Pontian. From Asia alone, upwards of fifty genera have been recorded in the Pontian, and although future study will doubtless show that some of these have to be united, the variety remains quite staggering. In Europe more than twenty genera have been recorded, while about a dozen are known from Africa.

The most primitive bovids are represented by the gazelles and related

Figure 39. With the *Hipparion* fauna in the early and middle Pliocene came the great evolutionary burst of the antelopes, which is still going on. The plains of Europe, Asia and Africa swarmed with great herds of animals like the *Palaeoreas* shown here.

Figure 40. The giraffid *Helladotherium*, first found at Pikermi, Greece, was the largest ruminant of the European *Hipparion* fauna, but was a comparatively short-necked form resembling the present-day okapis rather than the true giraffes. It was also in North Africa and southwestern Asia.

antelopes. Some of the Pontian gazelles were brachydont woodland forms, doubtless a primitive condition, but the majority were hypsodont and plains-living. Another primitive antelope group is represented by *Tragocerus*, in which the horns are relatively short and stout, and almost straight or slightly recurved. Many related forms are present in Eurasia; the group apparently survives in the present-day nilgai. There are also numerous forms related to the so-called hippotragines (oryx, addax and sable antelopes) and a few relatives of the reduncines (waterbuck, kob, etc.). Finally there is a series of forms with spirally twisted horns, related to the kudus and elands; this group is especially well represented in the eastern Mediterranean area.

Apart from antelopes, there are forms that foreshadow the evolution of goats and sheep, and a large number of bovids that seem to be related to the musk oxen. Typical genera are the Asiatic *Urmiatherium* and *Plesiaddax*, both of which have very massive horns. In the latter, the males are very much larger and have much heavier horns than the females; it is an extreme case of sexual dimorphism. Also possibly related to musk oxen is the remarkable *Tsaidamotherium* of central Asia with its asymmetric horns: the right horn, perched almost centrally on top of the head, was enormously broad and massive, while the left was only a small knob.

In North America the place of the bovids and giraffids was taken by the extravagantly-horned *Cranioceras* and *Synthetoceras* groups, and particularly by the prongbuck family, of which about nine genera are known in the Pliocene. These include immediate predecessors of the living *Antilocapra* as well as forms with forked or twisted horn-cores, which must be well off the modern line. The horn-core of the present-day *Antilocapra* is unbranched and the prong at the end of the horn is deciduous; the arrangement thus differs in principle from the forked horn-core of the Pliocene *Sphenophalos*.

The tylopod group of ruminants – camels and oreodonts – were still present in North America in the Pliocene, but the oreodonts were now very rare and only one genus survived to the late Pliocene – the small, graceful grassland form *Merychyus*. The camels, on the other hand, prospered; besides forms ancestral to modern camels and llamas, there were several side branches like the giraffe-camel *Alticamelus* with elongated neck and limbs, and the gazelle-camel *Rakomylus*.

Turning to the non-ruminant artiodactyls, we may note that the anthracotheres were now gone in North America and Europe, but present in Africa and Asia. The last of all the anthracotheres was the flat-skulled, long-jawed *Merycopotamus* which survived into Pleistocene times in southern Asia. Instead, the first true hippopotami make their appearance in the later Pontian of Europe.

The Pliocene record of the peccaries is unfortunately incomplete, only one genus being known; in contrast, the Old World swine are profusely varied and common in the Pliocene. Besides Miocene survivors like the listriodonts, we find true swine like the ubiquitous *Sus erymanthius* of the Pontian, water hogs (*Potamochoerus*) and many extinct forms. Among the latter may be noted the 'horse-pig' *Hippohyus*, with very high-crowned cheek teeth; the gigantic *Tetraconodon*, a tremendous creature reminiscent

Figure 41. The spoonbill mastodont *Gnathabelodon* may have used its thin-edged, shovel-like lower jaw to scoop up aquatic plants in shallow ponds. With a shoulder height of some eight feet, this was a medium-sized mastodont. Spoonbills and the related shovel-tuskers flourished in the Pliocene of North America.

of the ancient entelodonts; and the Chinese *Chleuastochoerus* or 'sneering pig' in which, in the males, the upper jaw forms a peculiar arch over the canines, with occasional 'cauliflower growth' of the bone. *Sanitherium* was a miniature suid with aberrant teeth.

The proboscidean lines of the Miocene continued to flourish in the Pliocene, but in addition there appeared various new forms, so that the diversity of the mastodonts was greatly increased. The main innovation in the Old World was the short-jawed *Anancus* group of mastodonts. An offshoot of the gomphothere family, they tended to reduce and finally lose the lower tusks, whereas the uppers became extremely long. In some late Pliocene and early Pleistocene specimens they may exceed ten feet. In the flesh this animal must have resembled the Pleistocene straight-tusked elephants though of course the cheek teeth retained their mastodont structure.

The *Anancus* group did not reach North America, which instead became an evolutionary centre for the highly bizarre shovel-tuskers. Although this group may have originated with primitive *Platybelodon* in Asia, its main deployment occurred in the Western hemisphere. They have the common character that the lower jaw and tusks formed a spade- or scoop-like structure, which evidently was used for digging, but there are several variations on this theme. Perhaps the weirdest looking is *Gnathabelodon* in which the lower tusks are absent and the mandible itself forms an enormously elongated spoon. It was presumably used as a mud digger, perhaps in foraging

Figure 42. Hoe-tuskers, *Deinotherium*, were probably mainly forest-living proboscideans. Their tall and long-legged build is striking; they culminate in the Pliocene with a shoulder height of nearly four metres (13 feet). Probably dependent on an equable, humid climate they were unable to cross the Bering bridge to the New World.

for aquatic plants, while shovel-tuskers like *Platybelodon* and *Amebelodon* may well have attacked firmer soil. In the former the 'shovel' is short and broad, while *Amebelodon* has a greatly elongated structure, the total length of the mandible and tusks being up to two metres.

Meanwhile the old mastodont lines – gomphotheres, rhynchotheres and the true *Mastodon* – continued with great success. An offshoot of *Gomp-*

hotherium, *Tetralophodon* tended to a more complicated molar structure, and the lower jaw became shorter. The aberrant *Deinotherium* with its hoe-like tusks reached greater size than ever before, culminating in the Pontian form which reached a height of four metres (13 feet) at the shoulders. The deinotheres, which were long-legged but short-necked and presumably had a long proboscis, belong to the woodland fauna.

Many other proboscideans preferred open country, and with the modernised ungulates they formed the great herds of herbivores which populated the Pliocene steppes and savannas. Together with them lived ostriches very similar to the living form; shells of ostrich eggs are not uncommon in Pontian deposits. Large land tortoises and monitor lizards are also common. The snakes are represented by cobras and vipers in Eurasia, rattlesnakes in North America; but the big pythons and boas are now gone from the northern continents – the last European pythons lived in the Miocene.

In the Pliocene, the ungulates of the Old and New World were strikingly different. In America we find horses, camelids, prongbucks, and various extinct horned and antlered types. In Eurasia, rhinos, *Hipparion*, deer, giraffids and bovids held sway. The similarity in the proboscidean faunas is slightly greater, and the same holds for the carnivores, although here too the effect of isolation can be sensed. The viverrid and hyaenid families, for instance, are restricted to the Old World. As regards the former this is hardly surprising, for the members of the civet family belong mainly to the tropical and subtropical zones and the climate of the Bering bridge would have been too cold for them. Both civets (*Viverra*) and mongooses (*Herpestes*) existed in Europe and Asia throughout the Pliocene.

The hyenas, which first showed up in the Miocene, really thrived in the Pontian; at most sites of this age, hyaenid remains outnumber those of all other carnivores taken together. Probably this is again a reflection of the very high productivity of the environment in terms of biomass. The family was much more varied than nowadays. The small *Ictitherium* was still rather like an overgrown civet, while the related *Palhyaena* comprises a number of species ranging in size from a fox to a wolf. The dentitions of these hyaenids suggest that they were highly predaceous. At several sites the remains of numerous individuals have been found together as if a group had been surprised by a flood. They probably hunted in packs, somewhat like the dholes and wolves of the present day.

There are also larger hyenas (*Percrocuta*), of which several species are known: the largest forms are lion-sized giants. Towards the end of the Pliocene, the earliest known members of the modern genus *Hyaena*

Figure 43. Larger hyaenids of the Miocene and Pliocene belong to the extinct genus *Percrocuta*, with members ranging in size from wolf to lion; the Chinese *P. gigantea* was the largest hyena of all times, while *P. eximia*, shown here with gazelle, was the typical hyena of the *Hipparion* fauna.

(striped and brown hyenas) made their appearance in Europe and Africa, and *Percrocuta* became extinct. The modern genus presumably evolved from a *Palhyaena* ancestor.

The cat family had undergone considerable modernisation and in the Pontian we meet with the present-day genus *Felis* for the first time. Its members are small, of tabby size, and the cats in the lynx-panther size range belong to the more primitive genus *Pseudaelurus*. In addition, there are various lines of sabre-toothed forms. The typical Pliocene sabre-tooth in both hemispheres is the large, powerful *Machairodus*. Earlier sabre-tooths possessed a large flange on the lower jaw, which acted as a sort of sheath for the enlarged upper canine when the jaws were closed. *Machairodus* had lost the flange so that the sabres protruded beneath the chin when the mouth was closed.

In North America there was also a quite recently discovered gigantic form retaining the flange; at the time of writing it is still nameless. Its canines are so inordinately enlarged as to remind one of a walrus, and the jaw flange was simply enormous. It is the most grotesque and aberrant felid known so far.

Among the mustelids we find martens, badgers, weasels, skunks and primitive wolverines (*Plesiogulo*) in both hemispheres, as well as various polecat-like forms. In Eurasia are relatives of the African honey badger –

a very large mustelid, and irascible and self-reliant in proportion. The otters are especially varied. Besides *Lutra*, the modern northern otter genus, there are several forms related to the clawless otters of Africa and southern Asia, and even to the sea otter of the Pacific. These otters have broader molar teeth than *Lutra* and their diet, instead of fish, may consist of crayfish, crabs, or molluscs; the broad flat teeth are used to crush the shells.

In the dog family, modern *Canis* evolved in the early Pliocene from the ancestral North American genus *Tomarctus*. The earliest representatives of true *Canis* are found in the later Pontian of Europe, but the dogs and wolves

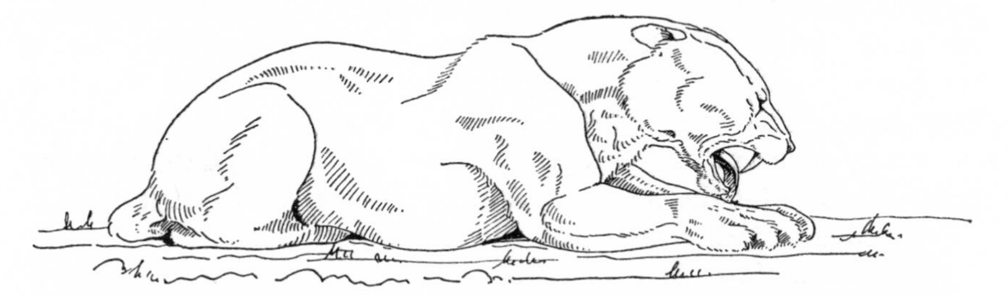

Figure 44. *Machairodus*, the most widespread sabre-tooth of the Pliocene grew to the size of a tiger or lion, was present in North America as well as the Old World.

did not become common until Pleistocene times. *Vulpes*, the true fox genus, also evolved in North America and reached Europe in the early Pleistocene. On the other hand, the raccoon-dogs (*Nyctereutes*) make their first appearance in the Astian of Europe. They are short-legged fox-like animals whose main history belongs to the Pleistocene and Recent.

Extinct dog types are still frequent in the Pliocene, continuing trends that became established in the Miocene or earlier. Thus the big *Amphicyon* is still found in the Pontian. The short-faced dog *Simocyon* is typical of the earlier Pliocene both in Eurasia and North America, and probably had hyena-like habits. It is succeeded in North America by the big hyena-like *Borophagus* in the late Pliocene.

Procyonids are rare in the Pliocene. The coati-like *Sivanasua* has been found in Asia, and the ringtails (*Bassariscus*) in North America. In the late Pliocene a panda-like animal, *Parailurus*, appears in Europe.

The great dog-bears are still found in the early Pliocene of North America (*Hemicyon*) and Europe (*Hemicyon*, *Dinocyon*) but they are now rare and their time is up. True bears are appearing instead. The largest forms belong to the genera *Agriotherium* and *Indarctos*. Some of these, especially the former genus, reached enormous size. As regards their anatomical structure, both forms are more than halfway between dogs and bears, but probably their habits were still more predaceous than those of most living bears. They were present in both the Old World and the New.

In the late Pliocene, the first members of the modern genus *Ursus* are found; they were no larger than the modern Malay bear. Recently, a small ancestral form, *Protursus*, has been discovered in the Vallesian of Spain.

By this time the ancient creodonts were long extinct almost everywhere in the world, but in India, curiously enough, we still find a member of the creodont family Hyaenodontidae, many million years after their demise in other continents. This is *Dissopsalis* of the Chinji, last of the creodonts. It was about the size of a leopard and resembled the early hyaenodonts of the Eocene, rather than the more specialised Oligocene forms. This is an example of the survival of the generalised, whose adaptation is so broad that minor environmental changes need have no adverse effect. The same principle holds for the opossum, which has survived although its more specialised marsupial cousins have become extinct.

At the same time, the survival of *Dissopsalis* in the early Pliocene of India is an example of how the southern continents furnish life room for organisms that have become extinct elsewhere. This does not necessarily mean that the tropical environment is less rigorous than, say, the temperate zone. It may simply be that the early Tertiary organisms of the northern continents were in fact adapted for life in a tropical environment and were unable to survive in other conditions.

The small mammals are not so well known in the early Pliocene as the large ones, but recent finds of fossiliferous fissure fillings, mainly from the later Pliocene, are now producing a wealth of information. Many of these fissures probably served as roosting places for owls and the fossil bones are the remains of their prey, mostly rodents and insectivores. Bat remains are common in many caves, probably from animals dying in hibernation. The bat cave floor, a soft mass mainly made up of the droppings of the bats, is an ideal environment for fossilisation.

In North America hedgehogs are still found in the early Pliocene – for instance the genus *Lantanotherium* common to Europe and North America – but after that we have no more traces of this family in the Western hemisphere. In Europe, on the other hand, hedgehogs continued to flourish. The same is true for the moles and shrews, and in Europe the amphibious desmans appeared in Pontian times; they are large, aberrant members of the mole family, adapted for swimming instead of digging.

The peculiar horned gophers or Mylagaulidae were still in existence in

North America, but became extinct at the end of the Pliocene. The squirrel family was highly varied in both hemispheres; there were true squirrels, flying squirrels, ground squirrels, and (in North America only) marmots and prairie dogs.

An important phenomenon is the explosive evolution of the voles, which got under way in the late Pliocene and led to a tremendous proliferation in the early Pleistocene. The stage was also being set for another, still later rodent explosion, that of the murid family (rats and mice); early members of this group were appearing in the Pliocene. The bipedal, jumping jerboas were common all over the Eurasian *Hipparion* steppes; they evidently arose in Asia, where their history goes back to the Oligocene. Their ecological counterparts in North America were – and are – the Zapodidae or jumping 'mice', which have a long but insufficiently known history going back to the late Eocene.

Dormice were still present in Europe in the Pliocene but most of the Miocene forms had become extinct and the varieties were much depleted. All the Pliocene forms belong to modern genera; of these two arose in the Miocene (*Glis* or fat dormice, and *Eliomys* or garden dormice), one in the Oligocene (*Dryomys*, tree dormice) and one in the Pliocene (*Muscardinus*, common dormice). The peculiar mole rats (Spalacidae) make their first appearance in the Pontian of Europe. They are highly specialised, blind, digging forms which are found only in eastern Europe and western Asia. Still another group, the bamboo rats (Rhizomyidae) evolved in Asia. They are also subterranean forms but use their teeth for digging, rather than the hands, as do the mole rats. They have also spread into Africa, where the earliest finds come from the early Pleistocene.

The beavers, which have an excellent fossil record, are common in the Pliocene. The dam-building modern genus *Castor* is present from the early Pliocene on in both hemispheres, but there are also many other genera. Some of them were related to *Castor* and probably resembled it in habits. Others belong to a quite different series of beavers, which probably were not dam-builders. They may have lived on aquatic plants, manatee fashion, whereas *Castor* eats land plants, especially the bark of aspen and sallow.

The two typical American families, pocket gophers (Geomyidae) and pocket 'mice' (Heteromyidae), were well represented in the Pliocene. Both groups possess large cheek pouches, but the pocket gophers are burrowing forms, while most heteromyids are hopping steppe and desert animals.

In the Old World we find porcupines (Hystricidae) both in Eurasia and Africa. This group may or may not be related to the New World porcupines, which evolved in South America. The two groups look rather different and differ in habits too, for the hystricids are excellent diggers, whereas the American porcupines are climbing forms.

In Pliocene times both the pikas (Ochotonidae) and hares (Leporidae) were common in all the northern continents. The origin of the true hares (genus *Lepus*) is disputed; both the American *Nekrolagus* and the Eurasian *Alilepus* have been cited as possible ancestors.

The marsupials had now vanished completely from the World Continent, as far as we know, although some forms may have survived in Central America.

The long separation of North and South America came to an end in the Pliocene when the Bolívar geosyncline finally became dry. (It runs parallel to the present-day coast of Colombia and Equador, from the Gulf of Urabá in the Caribbean to the Gulf of Guayaquil in the Pacific.) The mountains and dense jungle however continued to act as barriers (animals adapted to forest would be stopped by the mountains, and mountain-living animals would be stopped by forests) and intermigration was only a trickle, so that essentially the isolation of South America continued. However, some small mammals – cricetid rodents, for instance – managed not only to migrate from north to south but also to remigrate in the other direction after having evolved into new forms. In addition some larger mammals managed the journey, and so we find some intruders from the south in the Pliocene fauna of North America – ground sloths belonging to the families Megalonychidae and Megatheriidae, and, at a later date, glyptodonts, giant armoured forms related to the armadillos. These will be discussed in more detail in a later chapter.

The Pliocene fossil record of Africa is not very satisfactory. Generally speaking the fauna of northern Africa, which is the best known, is fairly similar to that of Europe and South Asia, whereas Africa south of Sahara is almost a blank for most of the Pliocene. We do find some remains of local forms such as the elephant shrews, the stegolophodont proto-elephants, and the hyraxes (of which only the modern genus *Procavia* was now in existence). The aardvark (*Orycteropus*), a typical African mammal of today, was also present in the Pliocene, not only in Africa but in Europe and southwest Asia as well. The big hippo-like anthracothere *Merycopotamus* is common in North Africa in the Pontian – another faunistic link with India.

Turning now to the Primates we may note that the richest Pliocene record is that of the Siwaliks and in Spain. There are also numerous other finds from Europe, both in the Pontian and Astian, and a few records from

Africa and East Asia. Prosimians are very rare, but a lorisid (*Indraloris*) is known from the Siwaliks, and we know that *Galago* was present in Africa since Miocene times.

The monkeys are well represented. Macaques appear in the early Pliocene in North Africa and spread to Europe and Asia in the late Pliocene. Baboons are present both in Africa and South Asia. The early langur *Mesopithecus* is the typical monkey of the *Hipparion* fauna in Europe and western Asia; in the Astian it was succeeded by a related form, *Dolichopithecus*.

Early gibbons (*Pliopithecus*) still existed in the Pontian forests of Europe, but they were gone in the Astian. The same is true for the ape *Dryopithecus*, which is especially common at the early Pontian site Can Llobateres, Spain. The extinction may well be ascribed to the effects of cooling. *Dryopithecus* is also found in Asia and an especially rich material has been collected in the Siwaliks. From time to time, fossil ape remains have been described under new names, so that a species list might give the impression of a veritable flood of apes in the Pontian of Eurasia. Recent revisions have made short shrift with those supposed genera – more than a dozen – and shown that all are really just *Dryopithecus*. There seem to be only two lines of evolution in that genus: a medium-sized one related to the chimpanzees, and a large, gorilla-like form.

The most interesting form, however, is doubtless *Ramapithecus*. It would appear that this early hominid spread from Africa into adjacent continents at the beginning of the Pliocene, for we now find its remains in the Pontian of Asia as well as Europe. The best material comes from the Siwaliks, where *Ramapithecus* is present in the Nagri zone. Some material has also been found in the coal deposits of Kaiyuan in the Chinese province of Yunnan, together with *Tetralophodon* and other mastodonts and an unidentified pig. This is probably also Pontian in date. On the other hand, there is no record of *Ramapithecus* from the rich *Hipparion* fauna of North China. In Europe, however, *Ramapithecus* occurs in the Pontian of Melchingen in the Swabian Jura. It seems fairly clear that all of these localities were situated in wooded areas, and that *Ramapithecus* did not inhabit the savannas of the typical *Hipparion* fauna. In its preference for woodlands it resembles the late Miocene *Oreopithecus,* which also had some man-like traits, although it was not a true hominid like *Ramapithecus*. The dependence on a forest biotope is evidently an important fact bearing on the early evolution of man. Unfortunately, it is very difficult to form a picture of the habits of *Ramapithecus* without knowing its postcranial anatomy.

In Africa the trail of the hominids is lost after the late Miocene of Fort Ternan and we only pick it up in the late Pliocene. At Omo near Lake Rudolf in southern Ethiopa a number of significant finds have recently been unearthed. Their radiometric age is upwards of four million years, corresponding to the Astian of Europe. These hominids are already typical *Australopithecus* closely allied to the Pleistocene ape-men which will be described in a later chapter. At Kanapoi in Kenya, fragments of a hominid have also been discovered, but at that time we are already at the threshold of the Pleistocene with its rich harvest of human fossils.

Let us finally take a look at the life of the Pliocene seas. Marine mammal-bearing beds are not so prolific as in the Miocene, but there are many scattered sites from which information may be pooled. In Europe, for instance, there are many records from the brackish inland sea of the south-east, for instance at Chisinau in Bessarabia and Kerch in the Crimea. The Mediterranean fauna is well represented in the Pliocene marine beds of Piedmont, Italy, and at Montpellier in southern France. In Belgium and Holland there is a good Mio-Pliocene sequence, and the Coralline Crag in East Anglia gives an interesting glimpse of the late Pliocene North Sea, where northern elements like the cod and the bowhead whale are already appearing in association with corals, molluscs of Mediterranean type, and tuna. There are also important marine localities in western North America, East Asia, South America, Australia and New Zealand.

The marine bird life was essentially modern in the Pliocene; the saw-toothed birds of the Miocene were now extinct, but albatrosses, pelicans, gulls, auks and many other forms are present, and remains of penguins are found in the south.

The desmostylids of the Pacific, which were described in the previous chapter, are still present in the earliest Pliocene but seem to have become extinct before the end of the Clarendonian. Early forms of the elephant seals appeared in the course of the Pliocene, and the walruses, previously confined to the Atlantic, invaded Pacific waters. The sea lions and true seals are essentially of modern type.

The dugongs, which form the great majority of the fossil sea cows, are today restricted to the Indian Ocean and nearby tropical waters, but in the Pliocene they were still found in the Mediterranean, the European inland seas, the Atlantic, and the northern Pacific. Dugongs are characterised by the presence of tusks in the jaw (in the males) and by the bilobed shape of the tail flukes. In the other group of sirenians, the manatees, the tail fin is rounded, tusks are lacking, and the cheek teeth are renewed in elephant fashion by replacement from behind. Pliocene manatees occur in South America and Florida.

Table 10. The Pliocene of the World Continent

Epoch	Africa (faunas)	Asia (faunas)		North America (ages)	Europe (ages)	Date (Million years)
		India	China			
Pleistocene	Olduvai	Pinjor	Nihowan	Blancan	Villafranchian	
	Omo	Tatrot				2
						3
Pliocene	Wadi Natroun	Dhok Pathan	Yushe	Rexroadian	Astian	
						5
			Paote	Hemphillian	Pikermian (Late Pontian)	
		Nagri				8
	Bled Douarah				Vallesian (Early Pontian)	
						10
Miocene		Chinji		Clarendonian	Tortonian	12
	Fort Ternan	Kamlial	Tung Gur	Barstovian		

Table 11. Number of mammalian families in the Pliocene of the World Continent

	Late Miocene	Pliocene	
		Early	Late
Total number of families	92	93	85
First appearances	5	2	1
Last appearances	1	9	2
Percentage newcomers	5½	2	1
Percentage extinctions	1	9½	2½

Among the whales a few extinct families are still found in the early Pliocene – the cetotheres, the squalodonts, and some long-jawed porpoises – but they are rare and the last became extinct in the middle Pliocene. Both the beaked whales and the dolphin family show a reduction in species since Miocene times. The Pliocene Delphinidae include some modern forms like true dolphins (*Delphinus*), killer whales (*Orcinus*) and long-beaked dolphins (*Steno*).

The sperm whale family was also less varied than in the Miocene, but there are still several genera in addition to the two that exist today (*Physeter* or true sperm whales, and *Kogia* or pygmy sperm whales). On the other hand, the whalebone whales of modern type were clearly increasing, as they replaced the vanishing cetotheres. The rorquals, humpback whales and bowhead whales were all present, together with extinct forms.

We are now at the end of the Tertiary period and at the threshold of the Ice Age. Our present-day world is still in the grip of that Ice Age; we live now in an interglacial age, but that is only a brief interval in a long series of temperature oscillations. Looking back the Pliocene is something of a paradise lost, a climax of the Age of Mammals before the coming of the cold; a time when life was richer, more exuberant than ever before or after. A tabulation of the mammalian families shows this too; the changes are small and unimportant, but after the Pontian apogee some extinction begins to set in.

In that world, as we now know, near-human beings were already in existence. The late Pliocene hominids must already have been fully bipedal, judging from the anatomy of their early Pleistocene descendants. Was this true for the early Pliocene *Ramapithecus*, too? We may know before long, but at present he is to us little more than a face and a name.

Chapter 7

Australia

Behold, I have divided unto you by lot these nations that remain, to be an inheritance for your tribes.

Joshua, 23:4.

OUT OF the remnants of ancient Gondwanaland, Africa and India came into contact with the World Continent at a relatively early date, and their history has been treated in the preceding chapters. Now we must turn to the remaining chips of the old supercontinent: to Antarctica, Australia and South America, all of which remained isolated for most or all of the Age of Mammals.

We know very little of the history of Antarctica. Very likely, as it broke loose from Gondwana, it carried a fauna of archaic mammals and became the nucleus of an adaptive radiation, bringing forth its own version of a balanced fauna. Of this we know nothing. We do know that the early Tertiary seas around Graham Land (south of Cape Horn) were still fairly warm; as late as the Pliocene, marine animals from Cockburn Island indicate slightly warmer conditions than the present.

But whatever forms of life existed in early Tertiary Antarctica, their world must have become increasingly bleak as the Tertiary wore on. That much of Antarctica was glaciated in the late Pliocene appears quite certain, and there is also every reason to think that land ice was in existence in the Miocene, perhaps even earlier. We do not know when the last land mammals of Antarctica became extinct but it probably occurred well before the Pleistocene; the continent seems to have been a sanctuary for penguins and seals for a long time. A rich penguin fauna, dating from the late Oligocene or early Miocene, has in fact been discovered on Seymour Island (off Graham Land); there are at least four species, as well as some cetacean remains.

A great deal more is known about the history of Australia, even if land fauna is only known for the later half of the Cenozoic and really abundant material only in the Pleistocene. Recent paleontological field work has brought to light a number of Tertiary faunas, both in Australia and in New Guinea, and there is every hope of getting more.

Australia, as an ancient block, has been emergent throughout the Age of Mammals, except for marginal flooding of the western and southern coasts. In the early Tertiary, the climate was somewhat warmer than now, judging from the marine faunas; at that time Australia occupied a position considerably farther south than at the present time. The interior was humid, with several large lakes. Tasmania was connected to the continent by means of a land bridge, which was gradually constricted in the course of the Tertiary and finally flooded. The large Oligocene lakes were surrounded by beech woods with the southern beech *Nothofagus* as the dominant tree, but conifers are also common and predominate in the lignites of Oligocene age. The swamps probably carried a coniferous forest, while the surrounding hills were clothed in beech woods. A find of Tertiary resin from Allandale, Victoria, weighing thirty-four pounds, contained the leaf of a Kauri pine (*Agathis*), a millipede, a mite, a spider, and some beetles and ants. The assemblage is typical of a moist, wooded environment. So the aridity of the inner Australia is a late development, probably of Plio-Pleistocene date.

The climate gradually grew warmer, according to consistent evidence both from the marine faunas and paleotemperature analyses, culminating in the middle Tertiary. Meanwhile the lakes in the interior tended to dwindle and finally ceased to exist; the last were gone by the end of the Miocene. Heavy volcanic activity occurred in the middle Tertiary, and in Queensland gigantic lava flows filled the valleys and formed great plateaux.

While the internal basins filled up and the lakes vanished, the seas encroached upon the southern and western coasts. One deep embayment east of Adelaide extended into the southwestern part of New South Wales, while another gulf covered the Nullarbor Plain as an extension of the Great Australian Bight.

In the middle Tertiary all of Australia was tropical or subtropical. The *Araucaria* (monkey-puzzle tree) then grew in Tasmania and Victoria; today it is restricted to New Guinea and the coastal rain forests of Queensland and northern New South Wales. The closely related *Agathis* or Kauri pine, similarly a tropical form and not found south of Queensland at present, has also been identified in the Tertiary of Victoria and South Australia. Crocodiles are now restricted to the northern part of Australia (where Rockhampton on the east coast, and Collier Bay in the northwest, represent their southern extremity); but there are Miocene and Pliocene finds as far south as Victoria. Oligocene and Miocene limestones in northern Tasmania contain corals, sea urchins, and molluscs (cowry, purpura, etc.) of tropical or subtropical type.

Towards the end of the Miocene the geography of Australia was greatly affected by faulting and uplifting; this caused the seas to withdraw for the time being from the margins of the continent. There was a less extensive transgression in the Pliocene, but this was again followed by emergence. At the same time, the climate gradually became cooler. The limestones that were formed by the Miocene sea flooding the Nullarbor plain are still replete with the tests of tropical foraminifers. But in the late Miocene the

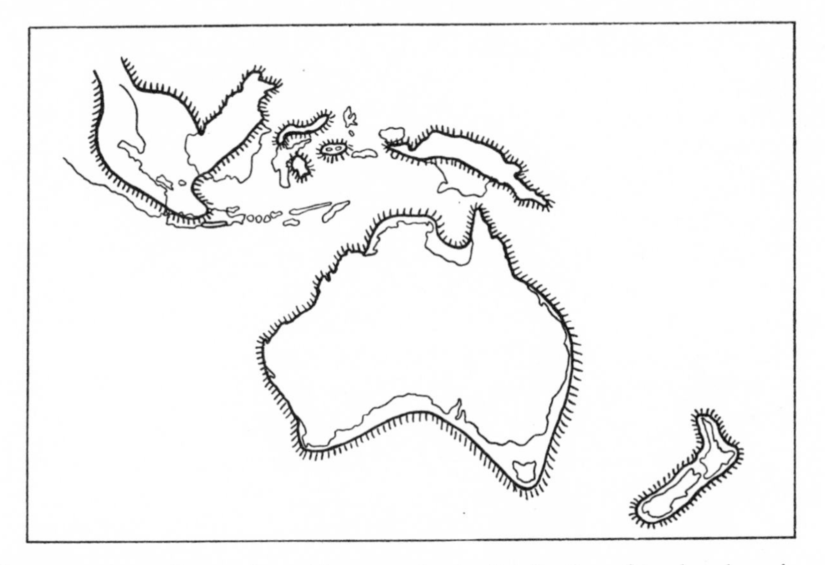

Figure 45. Australian region with approximate distribution of land and sea in the Miocene epoch.

tropical forms are becoming scarce, and the Pliocene marine faunas definitely reflect a cooling trend.

The history of New Zealand parallels that of Australia as far as climate is concerned. Here, too, a temperature rise in the early Tertiary is followed by an Oligo-Miocene maximum and a subsequent cooling. Much of what is now New Zealand was covered by sea in the early Tertiary and it was only the mountain-building phase of the Miocene that resulted in the fashioning and emergence of something resembling the present-day islands. In the late Cenozoic orogenic activity was resumed and added further to the area of the islands.

To the north of Australia lies New Guinea, which is connected with

Australia by means of the shallow Sahul Shelf; evidently it has been repeatedly in land contact with Australia and it forms part of the same zoogeographic region. But like New Zealand, much of New Guinea was a geosyncline in the early Tertiary, and its history is accordingly dominated by the geological turbulence of the Tethyan areas in the Cenozoic. Pushing up from the south and west, the Australian fragment of Gondwana threw up a bow wave in the shape of the Papuan geosyncline that now forms the mountainous backbone of New Guinea.

The early Tertiary marine formations of Australia and New Zealand give important glimpses of the vertebrate life of the southern seas. The large number of fossil penguins, especially in New Zealand, is particularly striking. Late Eocene deposits at Dunedin and Oamaru, Otago, carry a rich penguin fauna with at least half a dozen species. In addition, several late Oligocene forms have been found at Hakatara Valley in South Canterbury. These penguins are closely related to the fossil Seymour Island forms. In addition, whale remains, shark teeth, and fish otoliths (ear stones) are found.

In South Australia the late Eocene Murray River beds near Wellington also contain penguin fossils, and the same is true of the Gambier limestones from the late Oligocene; squalodont whales also occur at Mount Gambier.

Our first glimpse of freshwater life on the Australian continent is given by the Redbank Plains deposits of southern Queensland; they are thought to date from the late Eocene. In addition to a fossil flora, there are insects, fishes, crocodiles and turtles. The fish remains include the Australian lungfish *Epiceratodus,* a very ancient form. Another old-timer is *Phareodus,* a member of the tropical osteoglossids, which are among the most primitive living teleost fishes. Their range is now limited to the tropical freshwaters but they are descended from seagoing Cretaceous forms. Then there is a fish related to the 'milk fishes' of the Indian Ocean (*Gonorhynchus*), and an early freshwater bass, *Percalates.* The crocodilian belongs to the genus *Pallimnarchus,* which survived into the Pleistocene in Australia, and the turtle is a typical Australian long-necked turtle of the living genus *Chelodina.*

The Oligocene Table Cape beds of Fossil Bluff at Wynyard in northern Tasmania, though of marine origin (with teeth of the great white shark, *Carcharodon,* and remains of a squalodont whale), have yielded the earliest fossil of an Australian land mammal known to date. This mammal, *Wynyardia bassiana,* is a well-preserved skeleton that was found a century ago. Related to the living phalangers, cuscuses, and 'possums', it has a number of primitive characters which justify placing it in a family of its own, the Wynyardiidae. It was a creature the size of a small dog, quite powerfully-built and probably of a semi-erect stance somewhat like the living koala.

A more detailed history of land mammals in Australia begins with the late Oligocene or early Miocene Etadunna Formation of the Tirari desert east of Lake Eyre in South Australia. The formation consists of lake sediments which cover a wide area and have local concentrations of fossils. Near Lake Kanunka, for instance, many articulated skeletons of mammals were found in positions suggesting that they had been trapped in boggy clay. In addition to the various marsupial mammals, remains of many other vertebrates are found: lungfish, freshwater teleosts, turtles, crocodiles, monitor lizards, and many kinds of birds (pelicans, giant flamingoes, ducks, cranes and gulls).

At Lake Ngapakaldi river channels cutting into the Etadunna Formation have yielded a somewhat younger marsupial assemblage, and a still younger (probably early Pliocene) assemblage comes from the Mampuwordu Sands, which are stream channel deposits exposed at Lake Palankarinna. Crocodiles, a small emu, and various marsupials have been found here.

A rich early Pleistocene fauna occurs at Lake Kanunka in the Katipiri Sands which consist of stream and floodplain deposits. Again we find lungfish, freshwater teleosts, turtles, monitors and crocodiles. The birds include cormorants, ducks and swans, and also some large flightless forms related to the emus and cassowaries but belonging to the extinct dromornithid family. Several species of marsupials are found here, and in addition one true (placental) rodent. (See Plate 8.)

The record of the Tirari desert extends to the late Pleistocene, for the locality termed Cooper Creek has yielded a fauna of this age. These fossil bones were known to the aborigines inhabiting the area, who had legends about them. The fish and reptiles are in general similar to those from Lake Kanunka, and some of the birds also belong to the same types as in the earlier fauna. But this rich bird fauna includes many more species, such as geese, eagles, cranes and owls. In the rich mammalian fauna, both rodents and marsupials are represented.

In addition to the long sequence of the Tirari desert, fossils found at other localities both in Australia and New Guinea fill out the picture and give us the faunal history of this area in the late Tertiary and Pleistocene. At Alcoota in the Northern Territory a rich concentration of vertebrate fossils has been found, probably dating from the late Miocene. The Sandringham Sands, Victoria, contain an assemblage that is thought to be early Pliocene in age; these beds are marine but also contain land mammals. A good Pliocene fauna is known from Otibanda, New Guinea, where potassium-argon dates of 5·7 and 7·6 million years show that the deposits are approximately mid-Pliocene. The fauna is closely similar to the marsupials of the Australian mainland and indicates that the connection between Australia and New Guinea is of long standing. Besides the usual run of crocodilians and marsupials, the material includes remains of a boa-like snake, a cassowary, and a murid rodent.

Rich Pleistocene faunas have long been known from Lake Calabonna in southeastern Australia, the Darling Downs in eastern Queensland and numerous other localities, both caves and open-air sites. Recent exploration has added much to our knowledge.

We can only speculate about the way in which Australia was originally populated by the ancestors of its Tertiary and living fauna. The most primitive of its mammals, and indeed the most primitive of all living mammals, are the egg-laying monotremes – the duckbill platypus and the spiny anteater. These might represent a remnant of the original Gondwanaland populations of very lowly mammals in the Triassic and Jurassic, which became extinct everywhere else but was able to hold its own in isolated Australia. These animals, which are now highly specialised, probably adapted to their current ways of life very long ago, and did this so successfully that they were able to survive after the invasion of the higher or 'therian' mammals – the marsupials and placentals. There may have been other, less successful monotremes in Australia in those distant days, which were eliminated by the competing invaders. But, apart from some Pleistocene fossils which are practically identical with the living forms, we have no evidence on the history of the Monotremata.

If the monotremes represent the oldest 'faunal stratum' in Australia, the next is formed by the marsupials. Do they, too, represent a pre-rifting Gondwana population, or did they spread into Australia after its isolation? We do not know, but the latter alternative seems a little more probable. As far as we know, marsupials evolved only in the Cretaceous, at a time when Australia had probably become effectively separated from the other Gondwana continents. The first marsupials then perhaps entered Australia across the straits separating that continent from Africa or Antarctica, but probably not from Asia, across the immense expanse of the Tethys.

The early invader must have been a member of the order Marsupicarnivora, which is the only marsupial order distributed both within and outside Australia; living or fossil representatives are known from South America, North America and Europe, and the opossums of the Western hemisphere are typical representatives. The invasion probably occurred at some time in the later half of the Cretaceous, and the entry of a single gravid female, accidentally rafted on a floating tree, would be sufficient to populate the continent. Small opossum-like forms of this kind, ideally suited as structural ancestors of all the Australian marsupials, are in fact common in the later Cretaceous of the World Continent.

There is another reason, too, for assuming a waif dispersal across a rift strait. It appears that both placental and marsupial mammals evolved at about the same time from a common origin in the Pantotheria. Therefore, if Australia had been in land contact with the World Continent, it would seem difficult to understand why it was not invaded by placental mammals as well as marsupials. On the other hand, if dispersal was a question of a one-to-a-million chance across a marine strait, the absence of placental mammals is easy to understand.

The early *Wynyardia,* although a member of the order Diprotodonta, indicates that an origin from an opossum-like ancestor is highly probable, for it still retains some anatomical features suggestive of the Marsupicarnivora.

The third faunal stratum in Australia comprises the placental mammals. They entered the continent at a much later date, island-hopping from southeastern Asia; we shall return to them later on.

In present-day Australia and Tasmania, the Marsupicarnivora are represented by the Dasyuridae, the native 'cats' and 'wolves' of the region. The marsupial wolf *Thylacinus* is known from the Miocene on; in Pleistocene times it was still widespread in Australia, but now it survives only on Tasmania, perhaps because of competition from the dingo as well as pursuit by man. This animal is the size of a coyote, and not unlike a dog until you notice the peculiar kangaroo-like tail and the heavy, almost plantigrade hind feet. When pursuing its prey it runs like a dog, with great endurance; but if in danger it may take resource to a kangaroo-like, erect hopping. The Miocene forms are close to those of the present time and evidently lived in much the same way, preying on wallabies and other smaller marsupials; the Pleistocene thylacines tend to be larger.

While *Thylacinus* has been identified as early as the Miocene, the genus *Sarcophilus,* or 'Tasmanian devils', enters the record in the Pliocene. This animal, which has a head-and-body length of about seventy centimetres, is a very powerful, sturdily-built predator; it preys on wallabies, wombats, rodents, frogs, lizards, and so on. Fossil finds show that its range included the Australian continent up to the end of the Pleistocene, but it now survives only in Tasmania, like the thylacine. Pleistocene forms reach up to fifty per cent larger size than the living.

The marsupial 'cats' (*Dasyurus*) are smaller; however, a Pleistocene *Dasyurus* from the dune sands of King Island and Deal Island was much larger than the biggest of the living species. This animal appears to have been alive as late as 1801. *Dasyurus* species are good climbers and forage for birds and animals both in the eucalyptus trees and on the ground. The earliest *Dasyurus* known to date is from the Pleistocene, but the extinct genus *Glaucodon*, which is somewhat intermediate between *Dasyurus* and *Sarcophilus*, is known from the Pliocene at Ballarat.

The small order Peramelina comprises the single family Peramelidae, or bandicoots. Looking somewhat like peculiar, long-snouted rabbits, they are omnivorous in habits. They are known from the Pleistocene and Recent, but there is one form, *Ischnodon*, from the early Pliocene Mampuwordu Sands near Etadunna; it is closely related to the living rabbit bandicoot *Thylacomys*.

Table 12. Cenozoic land vertebrate localities in Australia and New Guinea

Epoch	Australia	New Guinea
Pleistocene	Lake Calabonna Cooper Creek Katipiri Sands	
Pliocene	Sandringham Sands Mampuwordu Sands	Otibanda
Miocene	Alcoota Lake Ngapakaldi Etadunna	
Oligocene	Wynyard	
Eocene	Redbank Plains	

By far the largest of the Australian orders is the Diprotodonta, comprising the phalangers, kangaroos, wombats, and a series of extinct groups, of which the Wynyardiidae has already been mentioned. The phalanger or cuscus family, Phalangeridae, makes its appearance in the Miocene with the Etadunna *Perikoala,* related to the modern koala. The true cuscuses, *Phalanger,* appear in the record in the Pliocene; they are now the most widespread of all the diprotodonts and the only marsupials to have spread into Celebes. The cuscuses are climbing forms; the first digit is opposable, both in the hand and the foot, and the tail is prehensile. These animals are therefore very apt to cling to the branches of a drifting tree and so be rafted across the sea. The family also contains various other kinds of marsupials including even a number of gliding forms somewhat resembling the flying squirrels of the World Continent. In the Pleistocene this family was quite diverse. There are some extinct forms, and the miniature, probably insectivorous *Burramys* from New South Wales and Victoria was long included in this group; but in 1966 the fossil amazingly came to life. A live *Burramys* was discovered at Mt Hotham, Victoria.

The small wombat family, Vombatidae, has a relatively short record. Only one genus is known as early as the Pliocene, the giant wombat *Phascolonus* which reached the size of a hog. Pleistocene remains of this great marsupial are plentiful, especially from Lake Calabonna. The wombats are the closest parallel to the rodents produced by the marsupials. The number of incisors has been reduced to one pair in the upper jaw and one pair in the lower; they are chisel-shaped and grow throughout life. Square-built and not unlike the koalas, the wombats differ by being excellent diggers rather than climbers. Grass, fungi and roots form their staple diet.

A still smaller family is the extinct Thylacoleonidae with the single genus

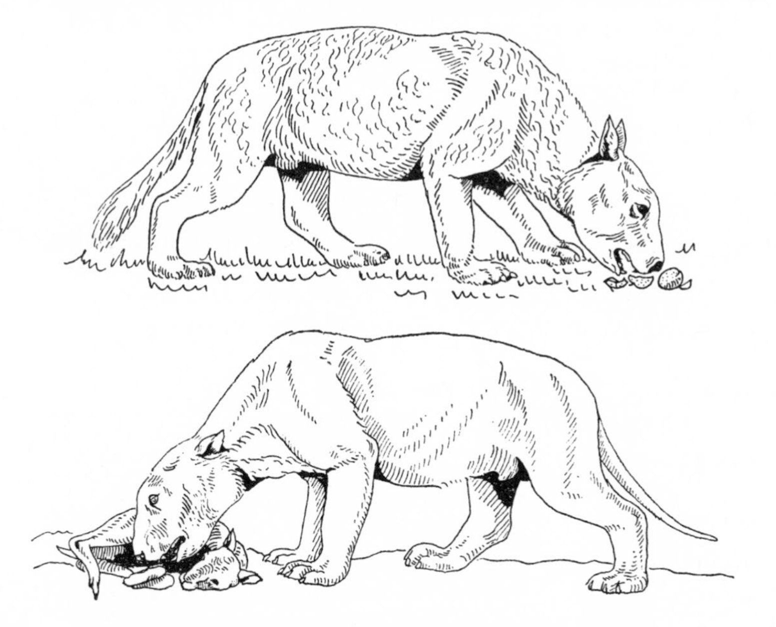

Figure 46. The 'marsupial lion' *Thylacoleo* has been regarded both as a harmless fruit-eater (top) or a carnivore (below, with killed wallaby). The animal was about four feet long, not counting the tail; Pliocene and Pleistocene of Australia.

Thylacoleo, of which one Pliocene and one Pleistocene species are known. Derived from the phalangerids, the marsupial lion reached almost the size of the true lion and possessed a most peculiar dentition. The middle incisors are large and may have taken over the role of the canines, for the true canine teeth are quite small. The most remarkable teeth are the enlarged premolars, one pair in the upper jaws and one in the lower. They are enormously elongated and form one pair of scissor blades on each side of the

Figure 47. *Sthenurus*, the browsing giant kangaroo of the Pliocene and Pleistocene in Australia. With their large size, heavy build and short tail, the sthenurines depart markedly from the main line of grazing kangaroos. Like so many other big mammals they became extinct at the end of the Pleistocene.

mouth. Nothing even remotely like this set of teeth is known in any living mammal and suggestions as to the mode of life of *Thylacoleo* range from fruit-eating to the most extreme rapacity. (It might be noted that one group of Cretaceous dinosaurs, the horned dinosaurs, had cheek teeth with a scissor-like action; they probably fed on highly fibrous vegetable matter, like palm and cycad fronds.)

The wallabies and kangaroos are grouped in the family Macropodidae, which also has a good fossil history and includes several extinct forms, among them some of gigantic size. Both the wallaby genus *Protemnodon* and the kangaroo genus *Macropus* have Pleistocene representatives exceeding the largest of the living kangaroos in size and reaching the height of a full-grown man. Even larger and heavier was the extinct *Sthenurus*, a big short-tailed kangaroo from the Pliocene and Pleistocene, which shows a secondary adaptation to a browsing mode of life. In contrast, *Macropus* is the marsupial equivalent of the horses and antelopes of the World Continent – the large grazing herbivores. The genus appears as early as the Miocene and its evolution may well be correlated with the swing in climate which seems to have set in by mid-Tertiary time, leading to a gradual desiccation; in Australia the interior lakes and forests were vanishing and grasslands, steppes and deserts were increasing, just as in other continents.

Figure 48. The largest of all marsupials, *Diprotodon* reached a length of about eleven feet. An inhabitant of the salt steppe, its remains are common in the Pleistocene of Australia; related forms range to New Guinea.

Of the other living genera of kangaroo-like forms, the New Guinea *Dorcopsis* dates back to the Pliocene at least, while the rock wallabies (*Petrogale*) and short-tailed wallabies (*Setonyx*) appear in the Pleistocene. The potoroos or rat kangaroos, which form a distinct division of the macropodid family, appear in the Miocene with the living 'squeaker' genus *Bettongia*; other genera appear in the Pliocene and Pleistocene.

The most spectacular group of Australian mammals is the family Diprotodontidae, which may be typified by *Diprotodon* the marsupial hippo, the largest known marsupial. Many skeletons of this great animal have been found at Lake Calabonna in southeastern Australia, where they were immersed in mud. Evidently the creatures were trapped and drowned here after having broken through the dried, salty crust that formed on the surface. In some instances remains of the stomach contents have been preserved and they show that *Diprotodon* fed on the typical plants of the saline plains (*Salsola* and other Chenopodiaceae).

Diprotodon represents the culmination of an evolutionary line within its family. There are several other genera and their history goes back to small, ancestral forms in the late Oligocene or early Miocene. A second evolutionary line within the family, represented by *Palorchestes* and related forms, tended to lightness of build (in fact *Palorchestes* was first thought to be a macropodid); this line, too, goes back to Etadunna days. Diprotodontids, except for the palorchestine group, are also found in New Guinea. About fifteen different genera of Diprotodontidae may be distinguished; of these, four are known in the Miocene, seven in the Pliocene, and seven in the Pleistocene. At the end of the Pleistocene, however, their downfall was swift and, like so many other large mammals, the diprotodontids became extinct.

Although we may tend to think of the Australian region as a stronghold of marsupials, it has also been inhabited by placental mammals since Tertiary times. Unfortunately next to nothing is known about these mammals prior to the Pleistocene and any conclusions about their history must be very tentative.

Bats have clearly been in Australia and New Guinea for a long time and their entrance presents no problem. The rodents are a different matter. All of the native rodents of the Australian region belong to the rat family, Muridae. They must have entered by way of the East Indian archipelago and New Guinea in a series of accidental rafting episodes – a mode of progress sometimes referred to as island-hopping.

The earliest invaders probably reached New Guinea in the Miocene, at the time when the Muridae of the World Continent were still in their infancy as a family. It appears that several different stocks entered the area at about this time, giving rise to the 'old Papuan' forms now mainly restricted to New Guinea. Two other ancient murine groups, the Hydromyinae (mainly of New Guinea) and Pseudomyinae (mainly of Australia) probably arrived in the Miocene too. Both have since multiplied and become highly diversified. For instance, the 'old Australian' or Pseudomyine group comprises six quite divergent adaptive types, most of which emulate World Continent forms, as follows: (1) unspecialised, like ordinary rats and mice; (2) medium-sized, short-tailed, slender-footed, resembling voles; (3) small, long-eared, long-tailed, saltorial, like the jerboas; (4) big, arboreal like the squirrels; (5) large-eyed, long-eared, hopping, like the rabbits; and (6) large, gregarious, nest-building – an almost unique adaptive type; obviously this group has had a long history in Australia.

The ancestors of the *Uromys* group probably entered in the Pliocene; this group seems to have arisen from something not very different from

Rattus. It first proliferated in New Guinea, then spread into Australia; this group still remains much more uniform than the descendants of the older immigrants. In the Pleistocene, East Indian members of the genus *Rattus* probably followed the same ancient trail and gave rise to several species now restricted to this area. Finally, prehistoric man introduced the dog, which became the dingo, and possibly also one species of rat; while modern man has been stocking both Australia and New Zealand – intentionally or unwittingly – with as incongruous a selection of creatures as may be imagined.

While the faunal histories of New Guinea and Australia (with Tasmania) have apparently been closely parallel, that of New Zealand is utterly different. No mammal seems ever to have reached these islands in geological time until the introduction of rats by prehistoric man. The vacant niches were invaded here by other animals, especially by ground-living birds. They are descended from flying ancestral forms, but most of them lost their powers of flight completely in adapting to ground life.

The main kinds of ground-living birds in New Zealand during the Pleistocene were the moas and kiwis, which belong to three different families: the short-legged moas (Emeidae), the true moas (Dinornithidae) and the kiwis (Apterygidae). All of these clearly were of very ancient origin in New Zealand.

Not much is known about the pre-Pleistocene history of the New Zealand land fauna, but the emeid moa *Anomalopteryx* is present in a late Miocene assemblage and thus indicates that the basic radiation of the moas is of very early date. The same genus is present in the Pleistocene and Postglacial. Several genera and species of medium-sized and small moa birds, including *Anomalopteryx,* were heavily-built and extremely thick-legged; they probably looked more or less like 'forty-gallon drums supported on knee-length gum boots'.

One of the smallest moas, *Megalapteryx,* evidently survived into Maori times in New Zealand, but most other forms were exterminated by the East Polynesian moa hunters that preceded the Maoris. Remains of *Megalapteryx* from dry caves in Otago District, South Island, are partly mummified with attached ligaments, skin, feathers, and eyes.

At Wairau on the northeast coast of South Island a moa-hunter camp has been excavated and has revealed the presence of great numbers of moa bones, as well as eggs perforated at one end to make water vessels. The commonest form here is *Euryapteryx,* a very stoutly-built, small moa with a very long neck and small head. Remains of *Euryapteryx* are also plentiful on North Island.

In contrast with the varied emeid family, the Dinornithidae comprise the single genus *Dinornis* or the true moas. In spite of their great height, ten-twelve feet in the tallest forms, these giraffe-birds were lightly built and slender; and their eggs, though large (up to four litres) were only half the size of the Malagasy elephant-bird eggs. Yet these moas are the tallest birds known.

Skeletal remains of *Dinornis* have been found in almost incredible profusion in a swamp in Pyramid Valley, where the alkaline peat is estimated to contain some 800 individuals of this and other genera per acre. The peat is now covered by a fairly tough crust, but at the time of the moas this

Figure 49. Giraffe-bird *Dinornis*, largest of New Zealand's extinct moas, was still in existence in historical times and was exterminated by man. Three metres tall or more, it was the largest known bird; runners-up include the thunderbirds of South America, the terror cranes of North America and Europe (see figure 7), the elephant birds of Madagascar and the living African ostrich.

was evidently thinner. The giant birds, lured into the swampy forest to feed, would break through and be caught helpless. As they sank struggling, they were often attacked by the great eagle, *Harpagornis*, now extinct, which devoured the heads and necks of the mired birds. From time to time the eagles themselves were caught in the swamp. There are also the remains of many smaller birds, and a fair number belong to *Euryapteryx*.

The true moas may have preferred open country, like the emus, while the emeids were probably mainly forest forms like the cassowaries. The shape of the head and beak indicates that the moas were herbivorous and this is confirmed by subfossil moa droppings, which contain remains of plants only. Seeds and twigs have been found in the gizzard region of a *Dinornis* together with a bunch of gizzard stones weighing a total of five pounds. A radiocarbon date (see p. 195) of the woody meal remains shows that this moa was alive about AD 1300.

The only surviving family of the native New Zealand ground birds is that of the kiwi, named for its piercing cry that may be heard in the night. The kiwis are remarkable for their hair-like feathers. Their known history goes back only to the Pleistocene, but they may well have originated from the same group of flying birds that gave rise to the moas.

In addition to moas and kiwis there are various other flightless birds. The very rare takahe, *Notornis*, has recently been exterminated on North Island and now survives in a single valley on South Island. The weka rail is a forest-living flightless bird which preys on rodents, young birds and eggs, thus acting as a substitute for the carnivorous mammals. In contrast, the owl parrot is strictly vegetarian. This bird, which reaches the size of the great horned owl, still possesses functional wings, but their power is barely sufficient to break the fall when the bird jumps from a rock or branch to the ground.

The Pleistocene was ushered in by a marked change in the climate. In New Guinea and northern Australia tropical conditions still prevailed, but Tasmania was heavily glaciated during the coldest phases and the Australian interior seems to have dried out even more. Upper Pliocene beds at Hamilton, Victoria, record the transition from the typical conifer forests of the Tertiary to the modern Australian forest type which is heavily dominated by *Eucalyptus* and *Acacia*. However, even in the Pleistocene, interglacial climates appear to have been warmer than that of the present day: for Pleistocene crocodile finds are recorded in South Australia at the latitude of Port Augusta, far south of their present-day range.

To sum up, then, the land mammals of the Australian region result from a series of colonisations. The most ancient faunal stratum, perhaps representing a pre-rifting Gondwana stock, is that of the monotremes. The marsupial fauna arose through a late Cretaceous or early Tertiary invasion, while the rodents, of southeast Asian origin, have entered in a series of waves probably beginning in the Miocene. New Zealand, however, was not reached by any land mammals and was populated instead by a variety of birds that lost the power of flight.

Chapter 8

South America

The inhabitants of the plains and mountains, of the forests, marshes, and deserts, are linked together in so mysterious a manner, and are likewise linked to the extinct beings which formerly inhabitated the same continent.

Charles Darwin: *The Origin of Species.*

THE HISTORY of South America during the Age of Mammals is much better known than that of any other island continent. There is a great succession of fossil land faunas, beginning with the Paleocene. The story told is that of the deployment of a richly varied insular animal world, reaching its acme in the late Tertiary; then came the invasions of World Continent species and the slow but inevitable extinction of the local forms, with the exception of a few groups that were able to hold their own. It is a remarkable story, and especially striking is the close convergence between some of the animals of the Island Continent and unrelated but similar forms of the World Continent. Clearly the adaptive radiation in South America tended often to follow the same grooves as those of the World Continent.

The climatic history of South America confirms that there was a cooling trend during most of the Age of Mammals. Darwin, who collected warm-water molluscs in Tertiary sediments at latitude 50° S – and many such faunas are now known – noted this as evidence for higher temperatures than those of the present day. As regards equatorial South America, the Tertiary molluscs indicate that the climate has been tropical throughout the Age of Mammals.

Mountain-building began in the Andean geosyncline in the late Cretaceous, evidently synchronous with the Laramide revolution of North America; and, as in the Rocky Mountains, deposition during the early Tertiary took place in intermontane basins. To the western side of the growing Andes, troughs developed and the sea spread into these basins. In Miocene times a second great orogenic phase commenced and resulted in the folding of the whole Cordilleran belt, succeeded by a period of intense volcanism. As the orogeny abated, erosion quickly reduced large parts of the highlands to a level surface, especially in Peru: the Puna surface. Towards the end of the Tertiary the entire region was upwarped, with resulting dissection of the Puna. The picture is further complicated by faulting (Lake Titicaca is one of the results), and the uplift continued in the Pleistocene, during which epoch the Andes reached their present height.

Profound disturbances also affected the sea bottom near the Pacific coast, as shown by the tremendous depressions just off the coast of Peru and Chile. The difference in elevation from the Milne Edwards Trench off Peru to the Nevado Huascaran peak, 200 miles away, is no less than 42,000 feet.

East of the Cordillera the history of South America in the Age of Mammals was comparatively placid. The main Tertiary deposits are found in the Amazon Valley and in Argentina, and most of them are continental; but there is also evidence that embayments of the Atlantic occasionally encroached upon the east coast.

The Caribbean coast and the Antilles had a very complex history during the Tertiary. Volcanism was intense at times and the islands were built up gradually from the bottom of the Antillean geosyncline. Central America also originated as a string of islands with transient and shifting coast lines. In Plocene times the great Bolívar Trench separatıng Colombia from the Isthmus was finally laid dry and from this time on there was probably land contact between the two Americas, although intermigration at first was strictly limited by the barriers formed by alternate jungles and mountains.

That Central America really formed an outlier of North America in the Tertiary is definitely shown by the presence, in the Miocene, of typical North American mammals in what is now the Panama Canal zone. This fauna, which seems to date from the beginning of the middle Miocene, comprises oreodonts, protoceratid ruminants, the paired-horned rhino *Diceratherium,* the forest horse *Anchitherium,* and others. This makes it clear that there was an unbroken land connexion to the north, and this is confirmed by another Northern assemblage in the early Pliocene of Honduras.

The only really rich Paleocene mammalian faunas outside North America come from South America and probably represent a fairly late phase in the Paleocene. In Patagonia the Rio Chico has yielded a mammalian fauna of this date. A very large, varied and still only partially studied mammalian assemblage of the same (Riochican) age comes out of ancient limestone fissure and cave deposits at Itaborai in Pernambuco, Brazil. These Paleocene mammals include marsupials, edentates, condylarths and notoungulates, which point to a relationship with the World Continent; all of these groups are also represented outside South America. In addition there are local groups, some of which are evidently derivable from condylarths, whereas others are more aberrant. No placental flesh-eaters are found in this assemblage and this resulted in the massive entrance of marsupials into this niche. Henceforth, and up to the invasion of the World

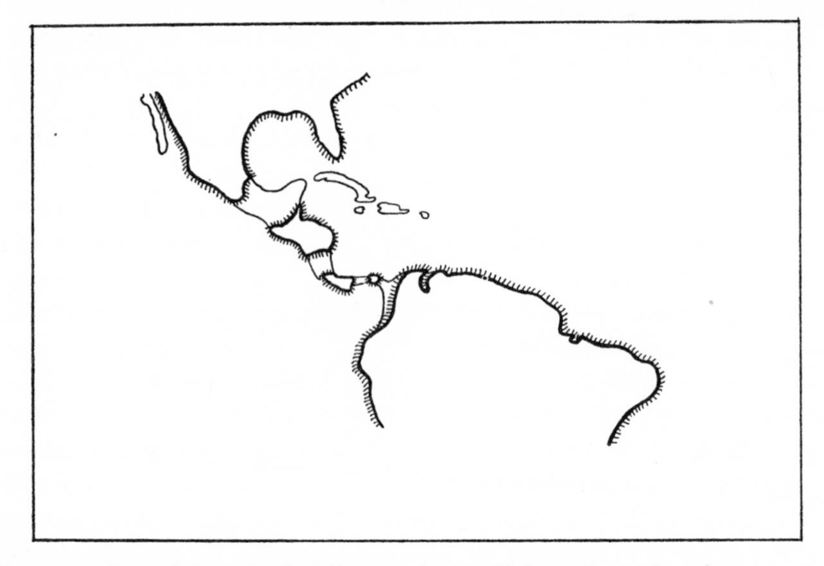

Figure 50. Central America in Miocene times still formed a series of stepping-stones between North and South America. The main barrier was that formed by the Bolívar Trench at the northwestern corner of the South American continent.

Continent families in the late Tertiary, the meat-eating in South America was handled by marsupials.

Early Eocene fossils are found in Patagonia in the so-called *Notostylops* beds of the Casa Mayor; the fauna is rich but consists almost exclusively of rather small forms. A slightly later stage, perhaps the middle Eocene, is represented by the *Astraponotus* fauna of the Musters beds, also in Patagonia. Late Eocene material comes from the Divisadero Largo in Argentina.

Marsupials in great variety are present in the Eocene; primitive armadillos and glyptodonts carry on the history of the Edentata. At the same time the condylarths are being overshadowed by the typical South American ungulates. Interesting reptilian forms include a group of very peculiar crocodiles, the Sebecosuchia, and the remarkable horned turtle *Crossochelys*.

The Oligocene starts off with the Deseadan age, named after Deseado, Argentina, with its *Pyrotherium* fauna. This represents a peak in the earlier evolution of the South American fauna, in which many forms reached great size. The indigenous, typical South American orders are now completely dominant, and the marsupial carnivores are varied and striking. Also found are the first of the giant flightless seriemas, or thunderbirds, which played a prominent role in the later Tertiary of South America. Finally, the first true rodents known from South America make their appearance in the Deseadan; they are the first invaders from abroad since the earliest Tertiary.

A late Oligocene mammalian assemblage, the *Colpodon* fauna, comes from the Colhué Huapi beds of Patagonia. Here we find the earliest records of yet another invading group, the Primates, in the form of South American monkeys.

We now come to the early Miocene or Santacruzian age in the South American succession. It is named after the Santa Cruz beds of Patagonia, which have yielded the richest of all the Tertiary assemblages of this continent. Many excellently preserved skeletons have been recovered. The edentates now form a large proportion of the fauna and the native ungulates are present in great numbers; some of the orders are at their peak, while others are already on the wane. The marsupial carnivores are as abundant as ever; the rodents were increasing and there are a few records of monkeys. The rapacious thunderbirds are also on the scene. In contrast, the later Miocene or Friassian record, which comes from La Ventana in Colombia and the Frias River of Argentina, is less abundant. The last condylarths are present in the Friassian. At the same time we find the first land bridge invader from North America – a procyonid carnivore.

By and large, however, the South American fauna persisted throughout the Pliocene, unaffected by the appearance of invading forms. This epoch may be divided into an early (Mesopotamian or Entrerian) age and a late one (Montehermosan); both of the type areas, Entre Rios and Monte Hermoso, are in Argentina. The edentate lines had now deployed into a great number of very impressive groups; some of these animals reached enormous size. Of the typical South American ungulates only two orders remained in the Pliocene and Pleistocene, the Notoungulata and Litopterna. Towards the end of the epoch the first Northern ungulates entered the scene – some peccaries, a llama and a cervid; the vanguard of the great Quaternary invasion that was to transform the entire South American animal world. The thunderbirds continued to flourish and one of them even found its way to North America.

As in the Australian region there is a clear age stratification of the faunal elements. There is a very ancient stratum, consisting of the creatures that colonised South America in the late Cretaceous or early Paleocene; here belong the edentates, condylarths and other primitive ungulates, the marsupials, and among non-mammals such forms as the sebecosuchid crocodiles, the horned turtles and the seriema birds. A second stratum is formed by the mid-Tertiary invaders, the caviomorph rodents and the monkeys, which arrived before the development of a continuous land bridge to the north. Third come the late Tertiary and Pleistocene invaders across the land bridge, at first only a trickle, then a steadily increasing stream.

An incidental delight in the study of South American fossil mammals is the peculiar system of scientific naming initiated by the pioneer student of Argentinian fossil mammals, Florentino Ameghino. Some of the names are in honour of contemporary scientists, usually with Hispanicised first names – *Ernestohaeckelia,* a litoptern (named for Ernst Haeckel the German evolutionist); *Edvardotrouessartia,* a notoungulate (named for the French paleontologist Edouard Trouessart); *Asmithwoodwardia,* a condylarth (named for the British paleontologist A. Smith Woodward). Anagrams are popular as in the armadillos *Eutatus* and *Utaetus,* the litopterns *Macrauchenia* and *Cramauchenia,* the notoungulates *Hegetotherium* and *Ethegotherium, Toxodon* and *Xotodon.* Some of the names reflect Ameghino's belief that the South American animals resembled World Continent forms because of a real genetic relationship, and that in fact the South American mammals were ancestral to them. A fossil monkey for instance was named *Homunculus* to express the opinion that it was ancestral to man. Despite such mistakes, Ameghino's life work is one of the most remarkable in vertebrate paleontology.

The South American condylarths are grouped in a family of their own, the Didolodontidae, which resembled the phenacodonts of the World Continent; evidently the two were closely related. Most of the didolodonts have been found in Paleocene and Eocene strata. These early forms were generally small, but a large surviving representative has been found in the late Miocene.

There is good evidence of a transition from the Condylarthra to one of the typical South American ungulate orders, the Litopterna. They may well be compared to the Perissodactyla of the World Continent; as in perissodactyls, the number of toes tended to be reduced to three or even to one. The family Proterotheriidae, or horse-litopterns, of which the earliest forms were present in the Paleocene, ranks as the pseudo-horse family of South America; its last members lived in the Pleistocene. It reached a climax in the Miocene with remarkable counterparts of both the three-toed and one-toed horses. The three-toed *Diadiaphorus* was a pony-like, lightly-built pseudo-horse with small side toes resembling those of *Hipparion* and other three-toed plains horses. The one-toed *Thoatherium* is famous for out-horsing the real horses as regards the reduction of the

Figure 51. The ecology of *Macrauchenia*, the litoptern with nostrils on top of the head, is obscure. Artist's suggestion is that it was a wading animal, eating floating vegetation; position of nostrils makes it possible to go on browsing without breaking off to breathe. This is consistent with the strong walking legs and the long neck; the former speak against complete submergence, the latter against the presence of a trunk.

side toes; the tiny splint bones that remain as vestiges are even smaller than in *Equus*. *Thoatherium*, however, was much smaller than *Equus* and might perhaps more properly be compared with gazelle-horses like *Nannippus*. In their dentition the South American pseudo-horses lagged badly; their cheek teeth were low-crowned and of a rather primitive pattern.

A second family of litopterns, the Macraucheniidae, are typified by the Pliocene and Pleistocene *Macrauchenia*. It was a somewhat camel-like creature, but it has a very aberrant character in the position of its nostrils. The nasal opening has shifted backward on to the very top of the head, between the eyes. The evolution of this character can be seen in a sequence of fossil forms beginning with *Cramauchenia* of the Colhué Huapi, in which the opening is in its normal position.

It has been suggested that this peculiarity is due to the development of a proboscis, but the arrangement does not resemble that in elephants or tapirs very closely. It is more probable that *Macrauchenia* was an amphibious mammal, in which case it would be of advantage to have the nostrils on top of the head. Macrauchenia possessed fairly high-crowned cheek teeth, presumably for cropping some kind of swamp vegetation.

In earlier members of this family, proportions also tended to be llama- or camel-like, but these animals were smaller and lacked the specialised characters of *Macrauchenia* itself.

By far the greatest of the native ungulate orders was that of the Notoungulata, which as we have seen was also represented in North America and Asia. It soon became extinct outside South America, but within that island continent it radiated into a great number of families and formed the majority of the ungulates.

The most primitive notoungulates are grouped in the suborder Notioprogonia, which comprises a number of Paleocene and Eocene forms in South America, as well as the Asiatic and North American representatives of the order. These animals are small, with simply built molars; skeletal remains of a closely related form indicate that they were relatively short-limbed and probably rather slow-moving, even though the feet seem plainly to be digitigrade rather than plantigrade.

A large suborder of notoungulates, the Toxodontia, comprises no less than seven families, of which some are very rich in genera. Its earliest members appear in the late Paleocene Rio Chico and the suborder flourished throughout the Tertiary and well into Pleistocene times. All of the late survivors belonged to the family Toxodontidae; they were quite large animals, very heavily-built, with peculiarly short fore-legs. *Toxodon* in the flesh has been likened to a gigantic guinea pig. Amphibious, hippo-like habits have been attributed to this animal.

Earlier toxodonts were smaller and less specialised; a very small horn was present on the forehead of some types. The Toxodontidae arose in the Oligocene and reached their apogee in the Pliocene; the other families in the suborder reached their climax somewhat earlier, in the Eocene or Oligocene.

Among them may be noted the Homalodotheriidae, a rather small family with only four or five genera. These animals were quite large and of normal notoungulate build except for the forefeet, which carried powerful claws instead of hoofs. We are immediately reminded of the World Continent perissodactyls known as chalicotheres. Whatever the mode of life of these two groups, it seems highly probable that the homalodotheres were doing exactly the same as the chalicotheres in their own environment; it is one of the most striking examples of covergence known. In another family, the Notohippidae, the cheek teeth became rather complex and showed some resemblance to those of the horses.

The Leontiniidae is a small family, known only from the Oligocene. At least some of its members appear to have had a nasal horn somewhat like that of a rhinoceros. Of the bear-sized *Scarrittia* several complete skeletons are known; they show a massive, heavily-built animal walking on large feet with spreading digits.

Then there is a third suborder of notoungulates, the Typotheria. These

Figure 52. Remains of the great hippo-like notoungulate *Toxodon* were excavated by Charles Darwin; they are very common in the Pleistocene pampas formation of Argentine. Length about nine feet.

animals tended to become more or less rodent-like, and this is especially true of the very large *Typotherium* itself. It was the size of a bear; its large incisors were chisel-shaped and rootless, and the cheek teeth, which were separated from the front teeth by a toothless diastema, were also high-crowned and rootless. With this very progressive dental apparatus, however, went a primitive postcranial build with five-toed feet. Typotheres of this type flourished in the Pliocene and Pleistocene. In the related but much smaller interatheres the tooth row remained primitive and the incisors were of normal type. This family started in the Paleocene.

Another family of rodent-like notoungulates, the hegetotheres, range from the Eocene to the Pleistocene. *Hegetotherium* of the Oligocene aud Miocene was a sturdily-built creature the size of a small dog, with enlarged, gnawing incisors and well developed cheek teeth; there are the beginnings of a diastema but the teeth in this region are still present, though reduced.

Figure 53. With its short tail and probably leaping gait, the small typothere *Pachyrukhos* may well have been rather like a rabbit in the flesh. A member of the family Hegetotheriidae, it lived in South America in the Miocene and Pliocene.

In the related but smaller *Pachyrukhos* the specialisation had proceeded much further and this animal was very rodent-like indeed; its large eye sockets may suggest nocturnal habits. In life it may well have resembled a rabbit. Probably the invasion of true rodents tended gradually to force the smaller, rodent-like notoungulates out of existence; in the Pleistocene only some large types survived.

Besides the notoungulates and litopterns, there are a few small ungulate orders of somewhat uncertain relationship. The smallest of them all is Xenungulata, erected for a single genus, *Carodnia* of the late Paleocene. It was about the size of the North American pantodonts and uintatheres and, judging from the teeth, may in fact be distantly related to them.

In the Eocene and early Oligocene we find another order, the Pyrotheria, which culminated in the Deseadan (early Oligocene) *Pyrotherium*. It is a most remarkable animal, elephantine in size, and with cheek teeth resembling those of the mastodonts to an astonishing degree. The tusks are not unlike those of some early four-tusked mastodonts, and the build of the nasal region indicates that there was a trunk. This would be necessary because the neck was quite short. The great animal stood on massive, pillar-like legs.

The Eocene pyrotheres, beginning with the Casamayoran *Carolozittelia,* are smaller; so there was a story of gradual size increase in the pyrotheres. The history of the pseudo-elephants in South America, then, shows a very close parallel to that of the mastodonts in Africa during the time span from the early Oligocene to the early Miocene – but shifted, as it were, some fifteen or twenty million years backward in time.

The weirdest of all the South American mammals, however, were probably the astrapotheres, order Astrapotheria. They culminated in the Miocene Santa Cruz *Astrapotherium,* an animal the size of a rhinoceros. The upper incisors are lacking but the lower are well developed, as in the

Figure 54. The South American pseudo-mastodont *Pyrotherium* lived in the early Oligocene, was the largest and terminal member of its order, Pyrotheria. This animal stood about five to six feet high at the shoulders.

ruminants. The canines form great tusks, which grew throughout life. The upper tusks are quite long, while the lower were kept short by abrasion. There is a long diastema between the tusks and the large, high-crowned cheek teeth. The shape of the nasal region may suggest the presence of a proboscis, although this would seem a little doubtful because the animal had a fairly long neck. (See Plate 9.)

In contrast with the great size and heavy build of the head and neck, the spine and hindquarters are rather weak and the limbs seem undersized and surprisingly slender. The peculiarities of this creature, which seems to have been constructed out of spare parts from several unrelated animals, make it very hard to assess its mode of life; amphibious habits, however, have been suggested.

More primitive forms of the family Astrapotheriidae are known from earlier levels, going back to the Casamayoran, or early Eocene. In the Pliocene they were all extinct. The other family of the order, the Trigonostylopidae, comprises primitive forms in the late Paleocene and early Eocene.

Besides the native ungulates, the edentate order formed a major component in the South American fauna. This order is still prominent today, in contrast with the indigenous ungulates, all of which are extinct. Anteaters, tree sloths and armadillos are with us still and may give us clues to the appearance of those extinct edentates that played such a great role in the past.

The earliest known edentates are the palaeanodonts of the North American Paleocene, which have already been described in a previous chapter. They differed from the typical, later edentates mainly in the construction of the hinder vertebrae, in which there is a system of extra or 'xenarthrous' articulations between the vertebrae. A recent discovery of armadillo-like neck vertebrae in the middle Eocene of Wyoming indicates, however, that xenarthrous edentates may have been present in North America too in the early Tertiary.

In spite of the ordinal name most edentates do have teeth (the anteaters do not). But the teeth in the front of the jaws are generally absent and the cheek teeth are very simple in structure. Tooth enamel tends to vanish.

The first edentates to appear in the South American fossil record belong to the armadillo family, or Dasypodidae. Scutes of this group are already present in the Paleocene Rio Chico of Patagonia. In the Casa Mayor (early Eocene), many genera of armadillos are found; in some, like *Utaetus*, the teeth are still enamel-tipped. The Miocene Santa Cruz material is especially well preserved and shows us a variety of differently specialised armadillos, probably ranging from ant- to plant-eating.

In the late Tertiary, some armadillos tended to attain giant size and the late Pliocene and Pleistocene *Pampatherium* was as big as a rhinoceros. In comparison, the living giant armadillo *Priodontes* is of quite moderate size; still it may give us an idea of what its enormous extinct cousin looked like.

A few small groups of aberrant armadillo-like creatures are grouped in separate families, for instance the peculiar horned armadillos, *Peltephilus*, with a transverse pair of horns (formed by scutes of the head shield) in front of the eyes.

A second large group of armoured edentates appeared in the Eocene and after a modest beginning, embarked upon a great evolution in the Miocene and Pliocene. These are the glyptodonts, which must rank as the most completely armoured mammals of all times. In these mammalian tortoises there was a bony carapace covering the entire body, with a separate bony casque on top of the head and bony rings encasing the tail. In contrast, the armadillo protection consists of separate, articulating transverse bands. At first the glyptodont carapace was a mosaic of separate plates; in later forms these tended to fuse, at first in the shape of bands somewhat like those of armadillos, then in a single inflexible structure.

In Santa Cruz times the glyptodonts were still quite small and the carapace was made up of separate bands which permitted some movement. By the late Pliocene they had attained giant size, with lengths up to four

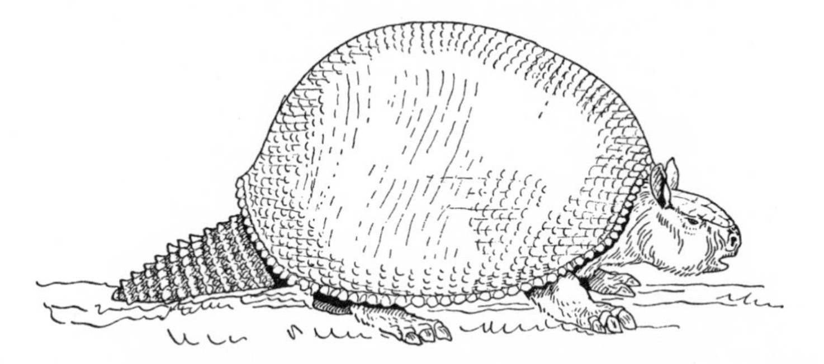

Figure 55. The Plio-Pleistocene *Glyptodon* and its relatives might be described as mammalian tortoises. Members of the order Edentata, they were related to the armadillos, but much larger (length about 9 feet). Glyptodonts originated in South America, ranged into southern North America.

metres or more. The tail might be covered by rings all along, as in *Glyptodon*; or its end part might be encased in a long tube-like sheath. In *Doedicurus* the tail ended in a structure resembling a spiked club, which must have been a terrific defensive weapon.

The armoured edentates just described form one great division of the order; another division is formed by the sloths and anteaters, which retain a normal hairy covering of the body. The anteaters, or Myrmecophagidae, are little known in the fossil state; there is one Santa Cruz form and the history of the living *Myrmecophaga* goes back to the Pliocene. Nothing at all is known of the history of the tree sloths, Bradypodidae.

The ground sloths, on the other hand, have an impressive fossil record. They are currently divided into three families. Their heyday came in the late Tertiary and Pleistocene, but early members appear in the Oligocene and there may be an ancestral form in the Casa Mayor early Eocene.

In the Miocene of Santa Cruz, *Hapalops* and related forms are the most common. These ground sloths are small, with lengths of up to four feet or so. They are fairly lightly-built, long of body and short-legged. Already they had the typical posture of the later ground sloths: the weight of the animal rested on the outer margin of the hind foot and on the outer knuckles of the hands, which were armed with great claws. A Pleistocene form, *Nothrotherium*, which spread into North America, seems to be a lineal descendant of *Hapalops*, but was about three times as big. It is also rather lightly built, and from preserved droppings is known to have lived on desert shrubs.

These sloths are usually grouped in the family Megatheriidae, which

Figure 56. With a length of six metres (20 feet), the South American ground sloth *Megatherium* (and the closely related North American *Eremotherium*) was the largest of the edentates. It was a treetop browser. Many edentates carry their young in this manner, but there is of course no proof that the ground sloth did.

also includes the gigantic leaf-eating megatheres which reached their climax in the Pleistocene. These animals were far larger than a modern elephant, being six metres long and very heavily built. They would sit up resting on their hind legs and stout tail to reach up to the foliage on which they browsed. Precursors from the Santa Cruz, although much smaller, show the megathere characters in an incipient stage.

A related but much smaller family, probably derived from early megatheriids, is that of the Megalonychidae. They first appear in Pliocene times and spread not only into North America but also into the Antilles. In contrast, their success in South America was indifferent.

Of greater importance were the Mylodontidae, a family of ground sloths which has a separate history going back to the Oligocene. There are sparse remains of a large mylodont in the Deseado early Oligocene, and more complete but smaller specimens in the Santa Cruz; the latter are no larger than a fox. Later mylodonts again tended to grow bigger, and one genus invaded North America.

The Pleistocene mylodonts were not as big as *Megatherium,* but much larger than *Nothrotherium.* This was the most abundant of the ground sloth families in the late Tertiary and Pleistocene. Dried patches of the skin of a mylodont from a cave in Patagonia show that the deeper layers retained innumerable small bone nodules; such nodules have not been found in the Megatheriidae or Megalonychidae. *Mylodon* itself appears to have been a grass-eating form.

We come now to the native carnivores of South America, marsupials belonging to the order Marsupicarnivora. Primitive opossum-like members of this order probably had an almost world-wide distribution in the late Cretaceous; but while they became extinct in the middle Tertiary in North America and Europe, they persisted in both South America and Australia and gave rise to a diversified host of descendants. In Australia, as has already been related, the radiation was a very broad one, leading to the occupation of a great number of different niches. In South America the presence of native ungulates prevented an invasion of that niche; but there appear to have been no competing carnivores, and so, with the characteristic opportunism of evolution, the marsupials proceeded to exploit that niche.

The South American carnivores are grouped in the family Borhyaenidae, which appears in the Paleocene Rio Chico. Most early boryhaenids were small, but in the Casa Mayor early Eocene appeared the bear-sized *Arminiheringia,* a powerful animal with very large canine teeth and the posterior molars developed as carnassials.

The Santa Cruz assemblage as usual is especially varied and well preserved; it includes the puma-like *Borhyaena,* as well as other forms which may have played roles analogous to those of coyotes, foxes and martens.

By far the most remarkable of the borhyaenids, however, is the Pliocene *Thylacosmilus,* or marsupial sabre-tooth. As in the sabre-toothed cats, the upper canines were modified into long, recurved, stabbing structures and the lower jaw carried a very large flange functioning as a sheath to the sabre. Details in the shape of the jaw and in the neck vertebrae show adaptive modifications very like those in felid counterparts. In both cases the sabre-tooth had to open the jaws very wide so as to free the points of the sabres; and furthermore, in both cases the neck had to adapt to the stabbing mode of attack.

The limbs are heavy and sturdy, as in some of the sabre-toothed cats, and the front paws had great freedom of movement and were evidently used in grasping the prey. In contrast with the felids, *Thylacosmilus* had no incisors at all, and the cheek teeth are of course of the typical marsupial pattern.

Except for this form the borhyaenids seem to have been on the wane in the Pliocene and all were extinct before the Pleistocene. Evidently this was

Figure 57. A marsupial carnivore, *Thylacosmilus* of the South American Pliocene shows an amazing resemblance to the sabre-tooths of the cat family. It is one of the last survivors of the family Borhyaenidae.

due to competition from the Northern Carnivora which were now coming in.

Besides the carnivorous marsupials, opossum-like forms (family Didelphidae) persisted throughout the Age of Mammals in South America; they are especially numerous in the Paleocene Rio Chico fauna.

A second South American marsupial order, the Paucituberculata, is also known in the fossil record since the Paleocene. It is represented today by the so-called opossum rats (Caenolestidae). They are small, rodent-like animals, which have actually been confused with rodents. Some forms have a dental battery very like that of the multituberculates; true multituberculates, however, have not been found in the Tertiary of South America.

A number of reptiles and birds were also important members of the ancient faunal stratum in South America and some of them merit special mention. There is, for instance, the remarkable horned turtle *Crossochelys* from the Eocene of Patagonia. It belongs to the very primitive amphichelydian turtles which were unable to draw the head into the shell. The horned turtles (family Meioloaniidae), which probably arose in the Cretaceous, were the last survivors of this very ancient turtle group. They persisted in Tertiary times in at least some of the Gondwana daughter continents. A gigantic Pleistocene form in Australia, *Meiolania,* was the last known representative of this group.

The aberrant sebecosuchid crocodilians, also found in the Eocene of Patagonia, are also survivors from the Cretaceous. The Eocene form is named *Sebecus.* Unlike other crocodiles they had rather deep, narrow jaws and head, and probably were more terrestrial in habits. Evidently they were highly predaceous and may have resembled the large monitor lizards.

That there was plenty of room for terrestrial predators in the South America of the marsupial carnivores is also shown by the rise of the thunderbirds, or giant seriemas. The present-day seriema is a mainly ground-living bird the size of a heron; besides large insects it will also prey on frogs and snakes, and may be regarded as a South American counterpart of the well-known African secretary bird. It is easy to see how birds living in this manner could have gone on to specialise on larger prey, as they themselves grew to larger size and lost all powers of flight; in due time this led to the evolution of the thunderbirds, or Phorusrhacidae, some of which were taller than a man.

The Miocene *Phorusracos* is the best-known of the thunderbirds. It was rather lightly built in spite of great size (five feet or more) and probably was very fleet of foot; it was armed with a great hooked beak. The related *Brontornis* was more heavily built. The first thunderbirds appeared in the early Oligocene and in South America the family persisted to the end of the Pliocene, then became extinct. But before that one member of the group was able to colonise North America; remains of this great predator, *Titanis,* have recently been unearthed in the Pleistocene of Florida.

The groups already described form the ancient stratum in South America. Next come the 'old island-hoppers', the caviomorph rodents and the monkeys, both of which entered by rafting across the marine straits from North America.

The first South American rodents appear in the early Oligocene, or Deseadan, but they are so highly diversified at that date – there are no less than six families – that they must have had a fairly long preceding history as yet unrecorded. Probably the introduction of ancestral rodents in the northern part of the continent dates well back into the Eocene. The Oligocene families are the Octodontidae (somewhat rat-like animals including the degus, chozcoris, etc.), the Echimyidae ('spiny rats'), the Chinchillidae, the Dasyproctidae (pacas), the Eocardiidae (an extinct family closely related to the guinea-pigs) and the Erethizontidae (American porcupines). True Caviidae, the guinea-pig family, appear in the late Miocene and the family had its heyday in the Pliocene.

In the Miocene appear the pacaranas or Dinomyidae, which radiated into a great number of Pliocene genera. One of these is *Eumegamys*, the largest rodent of all times; it was the size of a hippopotamus, which does seem incredible for a rodent. Unfortunately our knowledge of this monster is rather meagre. Various other big rodents belonged to the capybara family, or Hydrochoeridae, which arose in the Pliocene. Its sole living representative, the capybara, is the largest of present-day rodents.

Monkeys make their first appearance in the late Oligocene, or Colhuehuapian. There are two forms and they represent both of the living South American families: the marmosets or Callithricidae, and the Cebidae or typical South American monkeys. Such a division into families indicates that some time must have passed in this case, too, between the original introduction of the stock in South America and its first appearance in the record. The original invader certainly came from North America and may have been one of the Eocene prosimians, perhaps a member of the Omomyidae.

Marmosets are small, rather squirrel-like monkeys; apart from the Oligocene form and some Pleistocene finds of the living *Callithrix*, their history is unknown. The cebids are slightly better known; their history starts with the Oligocene and Miocene *Homunculus*, thought by its first describer to be ancestral to man. It is in fact related to the living owl monkey or dourocouli (*Aotus*); in contrast with most other monkeys, this is a nocturnal animal.

Table 13. Tertiary land-mammal faunas in South America

Epoch	Age (Locality)
Pliocene	Montehermosan Entrerian
Miocene	Friassian Santacrucian
Oligocene	Colhuehuapian Deseadan
Eocene	(Divisadero Largo) Mustersan Casamayoran
Paleocene	Riochican

In the late Miocene there are a few other finds, but the Pliocene is a blank and we have to go to the Pleistocene to find other fossil monkeys. All of these belong to living forms like the howlers (*Alouatta*), titis (*Callicebus*), miriki spider monkeys (*Brachyteles*) and capuchins (*Cebus*).

The third and youngest stratum of the South American land fauna is formed by the animals which immigrated over a land route from about the early Pliocene on. In the fossil faunas the first invader appears in the late Miocene. This is the procyonid carnivore *Cyonasua*. Bats were of course able to invade the continent at a fairly early date, but unfortunately their Tertiary history in South America is practically unknown. Peccaries, llamas and deer appear in the late Pliocene together with North American rodents of the 'New World rat' type.

The massive invasion did not come until Pleistocene times. As we leave the South American scene at the end of the Tertiary, it is still populated by an essentially native assemblage of mammals, where the local ungulates, the great edentates, and the marsupial carnivores form the most striking elements.

Looking back, we can see the evolution of a great, balanced animal world; in it we can perceive both unique forms of life (as in most of the edentates) and animals curiously resembling World Continent 'models': pseudo-horses, pseudo-mastodonts, pseudo-rhinos, pseudo-camels, pseudo-hippos, pseudo-chalicotheres, pseudo-sabre-tooths. Given the basic mammalian heritage, there seem to be certain adaptive types which have a tendency to develop; this is the important lesson of the Island Continent faunas.

Chapter 9

The Pleistocene Ice Age

> *They [the ice wedges] open, often with a sound and vibration as if from a distant explosion, in very severe frost, when the completely frozen soil contracts.*
>
> F. E. Zeuner: *The Pleistocene Period.*

LOOKING BACK on the Tertiary, we can perceive a number of trends in the gradual change of the face of the earth, its climates and its communities of living beings. There is an increase in the ruggedness of the earth's surface: mountains become higher, ocean bottoms deeper. The seas recede from the margins of the continents, so that the exposed land surface becomes larger. The climate becomes more diverse: in the middle latitudes the trend is towards cold, dryness and continentality. The continents become joined by land bridges and the land animals migrate from continent to continent; they increase their range, meet new competitors, do or die. The size of the animals tends to increase, especially in the mammals; there were more giants in the late Tertiary than in the early. Moreover, in all this it seems that a quickening of the tempo may be sensed.

The Pleistocene is a crescendo and a culmination of all these trends, and at the same time stands out as a major crisis in the history of life. Mountains are higher than ever, land areas wider, climates colder, animals larger. Climatic convulsions shake the earth; gigantic ice sheets turn immense tracts into frozen waste. Meanwhile early man proliferates and spreads to the end of the habitable world. At the same time the crisis mounts: the Age of Mammals is coming to an end; death and destruction succeed the exuberant Pliocene proliferation of life.

How did all this come about? The evidence now suggests that continental glaciation occurs only in conjunction with a long history of mountain-building. In the same way the orogenies of the Carboniferous and Permian periods, more than 300 million years ago, culminated in an Ice Age.

It is not difficult to see why this should be so. As the seas withdraw from the flanks of the continents, wide marginal areas are laid dry. Land does not retain solar radiation as efficiently as water, and so a greater amount of heat is radiated back into space and lost. Meanwhile the regression of the seas is accompanied by a corresponding lowering of the snowline, while great new highlands are pushed up to meet it. There is a great increase in the areas covered by permanent snow and ice.

Wherever the mountain ranges trend north and south, the air masses moving eastward are forced upward and cooled off, so that their moisture is precipitated in the form of snow. And so the amount of snowfall is increased at the cost of rainfall.

The snow that settles on the peaks above the snowline is gradually transformed by thawing and freezing into a granular substance termed névé, which is then slowly compacted by the weight of later snowfalls into ice. As the size of the mountain ice cap increases, there finally comes a moment when the weight of the overlying burden is so great that the ice at the bottom is squeezed out and begins to creep down the slope. The ice cap has now become a mountain glacier.

Now all depends on the balance between summer melting and winter snowfall. If the melting keeps pace with the addition of new snow, the glacier will persist in a steady state, neither growing nor shrinking. But if summer melting is reduced or winter snowfall increased, the glacier will grow larger. It becomes a river of ice down the mountain flank; it reaches the underlying lowland and spreads out into a piedmont glacier; it coalesces with other mountain glaciers to engulf the mountain range; and the piedmont grows into a great ice sheet.

At its maximum the ice sheet may accumulate up to a thickness of 10,000 feet. It is now a highland in itself and receives snow all over its expanse to keep the land ice going. The white surface reflects the sun's rays and robs the earth of heat. As water is bound up in the land ice, the sea level goes down correspondingly in a world wide regression of up to 300 or 400 feet. This also reduces the retentivity of solar heat. Everything is now playing into the hands of King Bore the monarch of the Arctic. It would seem that no end of continental glaciation is in view.

This is what happened in northern Eurasia and North America, and also – although to a lesser extent – in the southern hemisphere, wherever land masses extend far enough to the south, or mountain chains are high enough, to support a glaciation. Except for Antarctica, however, the largest ice sheets developed in the northern continents.

The greatest single ice sheet was the Laurentide of North America. This land ice, which finally covered upwards of five million square miles, had its core area around Hudson Bay and was thus exceptional in not developing on a highland nucleus. At its culmination it fused with the Cordilleran ice in the west, so that a single gigantic ice barrier stretched right across the continent from ocean to ocean. The southern margin of this immense land ice ran through the states of Washington, Montana, North and South Dakota, Iowa, Minnesota, Wisconsin, Illinois, Indiana, Ohio, Pennsylvania and New Jersey. That was the situation at the climax of the last, or Wisconsinian, glaciation; other end moraines of older date record slightly different conditions during earlier glaciations. South of the main ice sheet, smaller shields developed around the higher mountain ranges.

In Eurasia the greatest inland ice was that developed on the Scandinavian

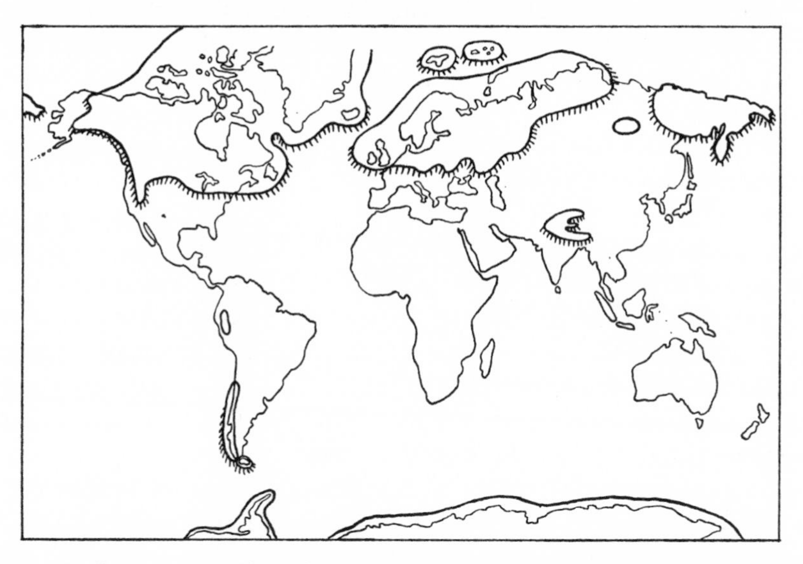

Figure 58. Greatest extension of continental ice sheets in the Pleistocene epoch.

mountains. In the west it coalesced with the ice fields of the British highlands, and in the east with the almost equally vast, but much thinner land ice based on the Urals. During the last, or Weichselian, glaciation in Europe the ice margin passed through the southern part of the British Isles, across the North Sea into Jutland, then south into northern Germany near Berlin, curving again gradually to the northeast to join the Ural ice in the White Sea region. The Alps and other mountain ranges had their own ice sheets.

South of the ice margin the land was a treeless, barren tundra much like that now found in the far north where the ground is permanently frozen. In sheltered valleys, willow and dwarf birch might grow; and in southern

France and southeastern Europe the land was partly forested. The climate in the Mediterranean area was cool-temperate at best, with pine forests of northern type.

In the southern continents, ice developed on the mountains of Africa, Himalayas, and even New Zealand; Patagonia carried a great piedmont glacier, and Tasmania was as heavily glaciated as the Alps.

But the ice does not last forever. There comes a time when melting begins to outpace the addition of new snow. The ice margin has already come to a halt and now the thickness of the ice sheet gradually diminishes. Then the ice margin slowly begins to withdraw; and the land surface, still depressed by the weight of the ice, becomes covered by large meltwater lakes.

As the climate becomes warmer, plants and animals migrate into the areas abandoned by the shrinking ice sheet. The hardiest, or those able to spread most rapidly, come first. The birch is usually the first forest tree to appear on a large scale; next a coniferous forest spreads into the area. If the climate goes on warming up, the area will presently be invaded by various kinds of broad-leaved trees, the so-called mixed oak forest. Meanwhile the ice dwindles and vanishes, or only a few mountain glaciers and ice caps are left. At that moment we are in the optimal stage of an interglacial.

The climate would be as warm as that of the present day, or warmer; many warmth-loving plants are recorded in interglacial beds far north of their present-day range, and animals like the hippopotamus and the macaque immigrated into the British Isles. The water stood higher, too, than at the present day; this is especially true for the earlier interglacials. As a result, in many areas the climate tended to have a more maritime character than now.

But the pendulum continues to swing and once more ice begins to accumulate. In the fossil record we can see how the vegetation belts, shifting southward, pass across a particular area. The mixed oak forest is succeeded by coniferous forest, and that in turn by birch; then comes the tundra vegetation, and finally – if we are within the area covered by land ice – the glacial drift deposits.

This is the pulse of climate in northern Eurasia and North America, and also in the smaller glaciated areas in most of the other continents. Antarctica, however, was certainly glaciated in interglacial as well as glacial intervals; in fact it probably carried inland ice as early as the Miocene, and other glaciations may have occurred in Arctic areas well back in the Pliocene.

On the other hand, tropical areas were little if at all affected by the cold. At most there seems to have been a change in humidity; in some areas rainfall tended to increase during certain glacial phases.

One major effect was the lowering of the sea level, which created land bridges where we now have shallow marginal seas. For instance, the Bering Strait was laid dry and formed a wide expanse of land, Beringia, over which cold-adapted land animals were able to migrate. The large East Indian islands Sumatra, Java and Borneo all lie on the Sunda Shelf which became dry land in the glacial intervals. At these times they formed part of the Asiatic mainland and only the Macassar Strait east of Borneo remained flooded. From here the way east, first to Celebes and then to the island world of New Guinea and Australia, was blocked by deep water. In Europe the English Channel and a large part of the North Sea were dry land and there was free access for land animals into the British Isles.

But while the seas receded, many lakes increased greatly in size and there were lakes where we now find only salt pans; even Death Valley in California had its own lake, Lake Manly, into which three separate rivers flowed. In Africa, Lake Chad was much larger than at the present time; in Europe, the Caspian Sea flooded wide areas of the plains of southern Russia.

Exactly how many times the swing between cold and warm has been enacted in the course of the Ice Age is still unknown. Usually, four separate glaciations are recognised both in North America and Europe; each has left its own record in the form of distinct glacial drift deposits separated by 'warm' interglacial beds with a record of temperate floras and faunas. In the area immediately surrounding the glaciated zone, the alternation may be one of loesses and buried soils. The loess was formed by windblown dust, which originated from frost weathering and was trapped by the steppe vegetation; the soil, on the other hand, is due to chemical weathering in a warm climate.

But it now seems clear that there were more than four glaciations. In the first place it is known that each of them may be subdivided into two or more cold phases; there are milder phases, termed interstadials, during which the ice sheets retreated and the climate became somewhat less rigorous, even if it did not get as warm as in the true interglacials. Secondly, there is good evidence of less extensive glaciations before the four 'classical' ones. Also, as we have already seen, there are traces of climatic oscillations of much the same type well back in the Pliocene and even the Miocene.

A paleotemperature curve for the Caribbean reveals no less than eight distinct cold phases during the last 425,000 years or so, giving an average period or wave-length of about 50,000 years. At least the two last phases represent the last (Weichselian or Wisconsinian) glaciation, while the earlier ones are at present difficult to identify with any certainty. Some continental glacial deposits have been dated by potassium-argon methods at about one million years and this is still only one-third of the total duration of the Pleistocene.

As we go back into the earlier parts of the Pleistocene, however, the evidence for continental glaciation becomes restricted to certain highland areas such as the Alps; and it seems clear that the land ice, even if it formed around high mountains and in the far north and south, was less extensive than in the later Pleistocene. As far as both the US and Europe are concerned, the early Pleistocene is pre-glacial.

The paleotemperature curve shows surprisingly good agreement with a curve based on the so-called perturbations of the earth's movement around the sun. Without going into detail as regards the perturbations, it can be shown that they affect the distribution of solar radiation received by the earth – although, of course, they have no effect on the total amount of radiation received. The effect will be a somewhat irregular oscillation between two extremes: (1) with comparatively cool summers and mild winters; (2) with relatively hot summers and cold winters.

The effects on the northern and southern hemispheres are somewhat different, although not antithetic as is sometimes mistakenly asserted. The effect on the northern middle and high latitudes is by far the most important because this is the zone with main glacial activity. Situation (1) with cool summers will reduce melting and promote the growth of inland ice, while (2) with hot summers will increase melting and lead to interglacial conditions. This seems at present to be the best available explanation of the peculiar cold-warm oscillation so typical of the late Tertiary and Pleistocene.

Table 14 summarises the most important climatic fluctuations that have so far been recognised in the Pleistocene history of Europe and North America. It has long been customary to keep 'Postglacial' time (roughly the last 10,000 years) as a separate epoch, the Holocene. But if we look at its climatic history, it has the unmistakable stamp of an interglacial.

As the Scandinavian land ice melted, for instance, northern Europe was gradually colonised by plants and animals coming in from the south, by way of Denmark, or from the east by way of Finland. In the early stages we find a tundra flora which takes its name after the beautiful mountain avens *Dryas*. This flora was succeeded in southern Scandinavia about 10,000 years ago by the invasion of birch, followed later on by pine, alder, aspen, hazel, oak, maple and elm. Some 8,000 years ago – a time when only large chunks of dead ice were left as remnants of the great land ice – the pine ranged far north, there were oak forests in the south, and the ash was appearing. As the warmth culminated some 7,000 years ago, leafy woods covered most of Scandinavia and pine forests extended north of the Arctic circle.

Then about 4,000 years ago, the optimum was passed and a cooling trend became evident. The spruce entered Scandinavia from the east and gradually spread southward along the Scandinavian peninsula. Hardwood forests dwindled and coniferous forests became dominant in almost all of Fennoscandia. Such is the situation at the moment. Despite slight temperature oscillations – like the short-range warming trend of the twentieth century – we are on our way towards a new glaciation. The Pleistocene is still going on.

The rapid swings in climate forced plants and animals to migrate in order to stay in a climate suitable to the adaptation of the species in question. In areas like North and South America, the main migration barriers – the mountain ranges – extend in a north-south direction; thus the plants and animals had little difficulty in moving equatorward or poleward. In Europe the situation is quite different. Here, not only the mountains but also the Mediterranean sea trend west to east. As the climate became colder and warm-adapted organisms were forced southward, they risked being caught in a cul-de-sac and becoming extinct.

Table 14. Glacials (GL) and interglacials (IGL) of the Pleistocene in Europe and North America

Epoch	Europe		North America	Date
Holocene	Flandrian	IGL		
				10,000
Late Pleistocene	Weichselian	GL	Wisconsinian	
	Eemian	IGL	Sangamonian	100,000
Middle Pleistocene	Saalian	GL	Illinoian	
	Holsteinian	IGL	Yarmouthian	250,000
	Elsterian	GL	Kansan	
	Cromerian	IGL	Aftonian ?	500,000
	Gunzian	GL	Nebraskan ?	
	Waalian	IGL	?	
				1,000,000
Early Pleistocene	Donau	GL	?	
	Tiglian	IGL	?	
	Essentially preglacial			
				3,000 000

Table 15. The Pleistocene of the World Continent

Epoch	South Africa (faunas)	Europe (ages or localities)	Asia (faunas)	North America (ages)	South America (ages)
Late Pleisto-cene	Florisbad-Vlakkraal	Late Pleisto-cene caves	'Loessic'	Ranchola-brean	
Middle Pleisto-cene	Vaal-Cornelia	Swanscombe, Steinheim	Yenching-kou	Irvingtonian	Pampean
	Swartkrans	Mosbachian	Choukou-tien		
Early Pleisto-cene	Sterkfon-tein	Villafranchian	Nihowan, Pinjor	Blancan	Chapad-malalian
Pliocene		Astian	Paote	Rexroadian	Monte-hermosan

The history of the European flora in the Pleistocene in fact shows a gradual extinction of 'exotic' elements. At the threshold of the Pleistocene, warm-climate forms such as the sweet gum (*Liquidambar*), swamp cypress (*Taxodium*) and tulip tree (*Liriodendron*) still formed about seventy-nine per cent of the flora in northwestern Europe. In later interglacials this percentage was gradually reduced, to nine per cent in the last interglacial about 100,000 years ago and zero in the present-day (or Flandrian) interglacial.

The animals fared somewhat better on the whole, but the continuous shifting of ground and the incessant stirring of the populations had a profound effect. The wide expansion of zones with an Arctic climate also affected the animal world and led to the evolution of numerous forms adapted to this kind of environment. It is clear that the pace of evolution, which had been rather slow in the Pliocene, now markedly increased. This also holds for the evolution of man.

The very complex glacial-interglacial stratigraphy is at present applicable in only a limited part of the world – mostly Europe and North America – and even here much is still provisional (see Table 14). In particular the correlation between the earlier glaciations in the Old World and the New is very uncertain.

Otherwise the practice is to divide the Pleistocene into three or perhaps four broad faunal ages, in analogy with the subdivision of the Tertiary epochs. Thus the 'preglacial' part of the Pleistocene is termed the

Villafranchian in Europe and the Blancan in North America. It would seem to cover roughly the time span between three and one million years ago; the highest radiometric age of a Villafranchian site (Mt Perrier in France) is 3·3 million years.

Several other sites in France, for instance Senèze and Saint-Vallier, have yielded rich faunas of Villafranchian age. At the former site, well-preserved skeletons of animals killed by volcanic gases and ash fall have been unearthed. The Arno Valley in northern Italy, and Villaroya in northern Spain, are other well-known find-spots. Most of the fossil remains from these localities are of large mammals, but numerous cave and fissure fillings in eastern Europe have yielded very rich material of small mammals, which has made it possible to study the faunal evolution in great detail; the Villány hills in Hungary are especially important.

In North America, Blanco and Cita Canyon in Texas are among the main localities of this age. The South American correlative is represented by the Chapadmalal beds of the southern pampas.

Nihowan in the province of Hopei and other Chinese localities have yielded animals of this age; in the Indian region, the Tatrot and Pinjor zones of the Siwalik series are equivalents of the Villafranchian. The famous sequence of Olduvai Gorge in Tanzania starts at slightly less than two million years and so represents the later half of the Villafranchian. In South Africa the ancient cave filling of Sterkfontein in the Transvaal is probably of about the same age. Both Olduvai and Sterkfontein have yielded fossils of primitive human types.

In South America the great Pampean formation contains the record of the later part of the Pleistocene. In the early nineteenth century, Charles Darwin excavated the skeletons of huge toxodonts and edentates in the Pampean and the obvious relationship between the fossil and recent edentates was one of the facts that led him towards the concept of organic evolution. At a slightly later date, the Danish explorer Peter Vilhelm Lund discovered richly fossiliferous caves of late Pleistocene date in Lagoa Santa in the state of Minas Geraes, Brazil.

Elsewhere, the middle and late Pleistocene may be subdivided into two or three ages. Many European sites date from the approximate time span beginning with the Waalian interglacial and ending with the Elster glaciation; this is the Mosbachian age, taking its name from one of the most prolific sites, Mosbach near Wiesbaden in northwestern Germany. Later middle Pleistocene sites like Swanscombe in England and Steinheim in Germany – both of which have yielded fossil human remains – date from the Holsteinian interglacial or the Saalian glaciation.

In North America all of the middle Pleistocene is regarded as a single land-mammal age, the Irvingtonian; the type locality is near Irvington, California. Some of the most important localities are found in southwestern Kansas, where a long sequence of Pleistocene beds forms a standard of comparison for the entire continent.

Northern China was populated by a rich mid-Pleistocene fauna, characterised by water buffalo and giant deer; we find its record at Choukoutien, the Peking man site. Further south the fauna is more tropical in character, with the proboscidean *Stegodon* and a giant tapir.

The Olduvai sequence of East Africa covers much of the middle Pleistocene, although there is probably a hiatus in the early part. In South Africa the early part of the middle Pleistocene, approximately equivalent to the Mosbachian of Europe, is represented by the local fauna at Swartkrans – another hominid-bearing cave filling in the Sterkfontein valley. Of somewhat later date are the remains of the Vaal-Cornelia phase, which have been unearthed in the terrace deposits of the Vaal River and in the Cornelia Beds of the Orange Free State.

Finally we come to the late Pleistocene, which in Europe comprises the Eemian interglacial and the Weichselian glaciation, and in North America the corresponding Sangamonian and Wisconsinian phases. This is the Rancholabrean age of North America, typified by the fossil fauna of the Los Angeles tar pits; there are also innumerable cave deposits of this age. The same is true for Eurasia, and in addition the loess deposits – especially in China – have yielded much information on late Pleistocene fauna. Among the many African sites of late Pleistocene age, the two thermal spring sites of Florisbad and Vlakkraal in the vicinity of Bloemfontein have been selected to typify it.

The radiocarbon method is used in dating the late Pleistocene and postglacial (or Flandrian). Radiocarbon, or C^{14}, is a radioactive carbon isotope which is constantly being produced in the upper atmosphere by the impact of cosmic radiation. This means that the atmospheric carbon dioxide will contain a certain percentage of radiocarbon atoms. It is assimilated by living organisms, which will thus contain a similar fraction of radiocarbon. When the organism dies, however, no new radiocarbon is taken up and the amount present will gradually diminish. By measuring the amount of radiocarbon left in organic tissue – bone, wood, frozen flesh or the like – we can determine its age; the method is practicable for ages up to 50–70,000 years.

As a result, we now have a fairly detailed chronology for the later part of the Weichselian glaciation. It has two main cold phases, separated by a fairly long (perhaps 10,000 years) interstadial with a cool-temperate climate. There were also numerous minor oscillations, with a final interstadial (the Allerød) beginning about 12,000 years ago. It was succeeded by the Younger Dryas phase (about 11,000 years ago) which was the last cold oscillation of the Weichselian glaciation. It ended about 10,000 years ago.

Most fossils of Tertiary land mammals come from flood-plain deposits, but in the Ice Age other modes of deposition are of special importance. The most spectacular of the Pleistocene fossils are of course the deep-frozen carcasses of animals such as mammoths, rhinoceroses and horses that have been discovered in the permanently frozen soil of Siberia and Alaska. They are the remains of animals that fell into fissures excavated by meltwater in the frozen subsoil, were killed by the fall, and then gradually refrigerated. However, all the carcasses were partly putrefied before they froze, and on thawing emitted a horrible stench. Some of the carcasses have been dated by the radiocarbon method and have yielded ages from 11,000 years to more than 44,000.

Another celebrated find is a pickled rhinoceros from Starunia, Galicia, which was preserved in brine and oil. Then there are the tar seeps of California, Peru, and the Caucasus; they have trapped innumerable animals and the asphalt may contain incredible numbers of bones. For instance, the remains of the great sabre-toothed cat *Smilodon* from the Rancho La Brea tar pits in Los Angeles must represent more than 2,000 individuals; and the big dire wolf *Canis dirus* occurs in even greater numbers. The pits would probably look innocent enough after rain, but the animals which came down to drink would get trapped by the treacherous tar beneath the water. The struggles of the mired animals would attract flesh-eating birds and mammals to the place; they would get trapped too and act as bait for further victims. Because of this chain-reaction, the bones of carnivorous species outnumber the vegetarians ten to one, or more. (Plate 10.)

However, the most important environment for fossilisation was that of the cave. Large caves were used as dens by Stone Age men or by carnivorous mammals; both would leave the bones of their prey, and occasionally their own bones, in the cave deposits. Bats, owls, or small carnivores would inhabit the smaller caves and fissures. The bear and hyena caves of the late Pleistocene in Europe are especially remarkable. Some caves contain the remains of several thousand cave bears and little else – up to ninety-nine per cent of the bones found may belong to this one species. It can be shown that the remains are of animals that died in their winter sleep, for the young individuals show a clear separation into newborn, one-year-olds, two-year-olds and so on.

A final source of knowledge of the extinct animals is formed by the paintings and engravings made by Ice Age man. The large game and large carnivores were most commonly portrayed: mammoths, horses, bisons, aurochs, lions, bears, and many other species are represented. The pictures are done with consummate artistry and a keen eye for the characteristics of the various species.

The beginning of the Pleistocene is marked by the appearance of three groups of mammals: the true elephants, the modern one-toed horses of the genus *Equus*, and the true bovines now represented by cattle, bison and buffaloes. The elephants and bovines evolved in the Old World and make their appearance in the New at a somewhat later date, while the horses are of course of American origin (descended from *Pliohippus*) and came into Eurasia by way of Beringia at the beginning of the Villafranchian. There are also various other forms that appear as migrants from continent to continent and thus are useful as Pleistocene index fossils.

The main difference between elephants and mastodonts lies in the structure of their cheek teeth. Those of the mastodonts have rounded cusps or transverse crests, which are comparatively few in number. In the elephants the crests are closer together and their number becomes greatly increased; furthermore the spaces betwen the crests, or lamellae, are filled with cement. The teeth thus become converted into huge grindstones in which the grinding surface is formed by three different tissues – enamel, dentine and cement, arranged in transverse bands. Since the three tissues differ in hardness (enamel being the hardest) the grinding surface is kept rough and efficient all the time by wear. As one of the great teeth is being worn down, a replacement tooth is developed in the jaw behind it and slowly pushes forwards to take its place.

Elephants also tend to have higher heads and shorter jaws, and are taller and slimmer of build than the mastodonts.

The elephants are descended from the stegolophodonts of the Miocene and Pliocene. A rather primitive type of elephant was *Stegodon*, which was common in Asia in the early and middle Pleistocene. In the males, the tusks were exceedingly large and emerged from the jaws so close to each other that there could have been no room for the trunk between them; it must have been held to one side of the tusks. The female had much shorter tusks with more space in between.

True elephants (*Elephas*) appear in the earliest Villafranchian in southern Asia and Africa, and shortly afterwards enter China; Europe was invaded at a somewhat later date, probably about two million years ago. These early elephants were still primitive in many respects but gave rise to a number of diverse and more highly evolved descendant types, which may be roughly divided into four groups.

In Africa there evolved a type (*Loxodonta*) mainly adapted to a savanna and open forest environment. It is fairly large, with very big ears to help in temperature regulation, and medium-sized tusks. It browses on leaves and succulent herbs, but will also eat bark (using the tusks and trunk to strip it off the trees), fruits and the like. Another type is the Indian elephant, which evolved in the jungles of southern Asia; it has more densely packed lamellae in the cheek teeth and the tusks are smaller, perhaps so as not to

Figure 59. *Stegodon*, a primitive Pleistocene elephant found in the Old World tropics and subtropics. Shown here is a male individual; in the females the tusks were shorter and more widely spaced.

impede the movements of the animals in dense jungle. Both of these species show a highly-developed social life, with devoted parental care of the offspring; this may be one of the reasons for their survival to the present day.

Two other groups of elephants evolved in the Northern continents: the straight-tusked forest elephants and the mammoths. The former are still rather like the early elephants, with widely-spaced lamellae in the cheek teeth, but they became even larger (up to four metres in shoulder height), and the tusks were nearly straight and of enormous size. This type of elephant (*Elephas namadicus*) inhabited the open forests and parklands of the Eurasian continent and was a characteristic member of the interglacial fauna in Europe. Foreign to the tundra, it was unable to cross Beringia and so remained confined to the Old World.

The mammoths form a separate lineage since Mosbachian times and evolved in adaptation to a diet based on tundra and taiga vegetation. The lamellae of their cheek teeth became more closely crowded than in any other elephants, and the late Pleistocene woolly mammoth lived on steppe grass and herbs as well as twigs and leaves of birch, osier and pine. This animal had a monumental appearance: immense, strongly curved tusks (which were used to scrape the snow off the grass), a peaked head and a humped, sloping back, and a thick, woolly coat of hair. In size, however, the woolly mammoth lagged far behind the gigantic mammoths of Mosbachian times, some of which reached a shoulder height of up to 4·5 metres; they were the largest known proboscideans.

There were several different invasions of mammoths in North America; here the animals became adapted to milder climates and colonised most of the continent. They did not cross the land bridge to South America, probably because they could not penetrate the jungle. Of the American forms the largest was the emperor mammoth (*Elephas imperator*), second only to the Mosbachian *E. trogontherii* of Eurasia. The late Pleistocene woolly mammoth is also found in North America.

Although the peculiar shovel-tuskers became extinct, some other types of mastodonts survived in the Pleistocene. *Mastodon* is found only in the earliest Pleistocene of Eurasia but survived into postglacial times in North America, where *Mastodon americanus* is one of the commonest proboscideans. Other mastodonts invaded South America; the peculiar *Cuvieronius* had spirally twisted tusks. The *Anancus* line of straight-tusked mastodonts persisted in the Villafranchian of Eurasia and Africa. In some of these creatures, the enormous tusks were as long as the body.

The ancient hoe-tuskers (*Deinotherium*) survived in the earlier Pleistocene of Africa, but in Eurasia their story came to an end with the Pliocene.

The horse group is dominated by one-toed horses in the Pleistocene, although three-toed horses persisted in some areas – the gazelle-horse *Nannippus* in North America, *Hipparion* in Eurasia and Africa, and even a recently discovered browsing horse (anchithere) in the Villafranchian of China. The one-toed forms, however, soon became completely dominant both in the Old World and the New. South America became the homeland of various small, short-legged mountain horses like *Hippidion*.

In Eurasia the earliest members of the genus *Equus* are very large, heavy horses somewhat like the present-day Clydesdales; they invaded Europe probably half a million years or so before the elephants. Later horses tend generally to become smaller and more graceful; in the late Villafranchian, donkey-like forms are already appearing, and in Asia the hemiones and onagers evolve. A branch of the Eurasian horse group seems to have re-entered North America in the mid-Pleistocene and rapidly ousted the indigenous forms.

Tapirs were common in Eurasia and both Americas during the Pleistocene, but they gradually died out in the northern continents so that only South America and southeastern Asia remain as tapir territory. The chalicotheres, or clawed perissodactyls, were also becoming extinct; the last of this group are found in the Pleistocene of Asia and Africa.

Besides the horses, the rhinoceroses were the only perissodactyls that continued to flourish in the Pleistocene, and they were by now an exclusively Old World group. Tropical forms are relatives or ancestors of the present-day rhinos of Africa and southern Asia. Northern Eurasia, however, was populated by a remarkable array of rhinoceroses, most of which were closely related to the living Sumatra rhino *Dicerorhinus*; as may be remembered, this is one of the oldest living mammalian genera, dating back to the Oligocene.

There are at least four species belonging to this genus in different phases of the Ice Age and living in different environments; the best-known is perhaps the so-called Merck's rhino so common in the European interglacials (*Dicerorhinus kirchbergensis*). The woolly rhinoceros (*Coelodonta*) is an aberrant offshoot of the same group; it evolved in Asia in the early Pleistocene and became a regular inhabitant of Europe during the Saalian and Weichselian glaciations. In spite of its obvious adaptation to a cold tundra environment, it does not appear to have crossed Beringia into North America.

This is also true for the only rhinoceros of northern Eurasia that was not related to *Dicerorhinus*: this is the gigantic *Elasmotherium*, an elephant-sized beast with a frontal 'horn' up to two metres in length. Its history is well known; it is a culmination of the line of giant rhinos that arose in Spain in the Miocene, and in the Pliocene was represented by the mighty *Sinotherium* of China. The elasmothere cheek teeth had a very complicated, almost horse-like enamel pattern, and were very high-crowned. In the Ice Age saga of the great Danish poet Johannes V. Jensen, *Elasmotherium* figures as a giant unicorn with a petrified heart.

The Artiodactyla reached the acme of their career in the Pleistocene. Peccaries in both Americas, and true swine in Eurasia and Africa, were plentiful; there are some remarkable giant forms like the great *Celebochoerus* which evolved in isolated Celebes, and the enormous African wart hog *Afrochoerus*, whose immense tusks were at first ascribed to a mastodont. Africa, in particular, was inhabited by a remarkable variety of suids related both to the wart hogs (*Phacochoerus*) and the water hogs (*Potamochoerus*).

Some of the ancient anthracotheres still survived locally in Asia and Africa; but generally speaking the hippopotami were now rapidly ousting these archaic forms. In Villafranchian times the hippos spread over Africa, southern Asia and Europe. The last indigenous British hippos lived in the Eemian interglacial, about 80,000 years ago.

The commonest peccaries in North America were the long-nosed forest peccary *Mylohyus,* especially in southern and eastern USA, and the flat-

Figure 60. The gigantic Pleistocene rhinoceros *Elasmotherium* carried a frontal horn up to two yards long. Of elephantine size, it was equipped with a very efficient grazing dentition; its homeland was the steppe of northern Eurasia.

headed peccary *Platygonus* which apparently preferred open ground and seems to have lived in large herds. Remains of perhaps a hundred *Platygonus* peccaries have been found in a cave in St Louis, Missouri; they may be the remains of a herd of these animals that was trapped in a fissure.

The camels had invaded South America in Pliocene times. Now, at the beginning of the Pleistocene, they crossed Beringia together with the one-toed horses and colonised Asia. While the South American invaders were llama-like, those that entered the Old World were ancestral to present-day true camels. They spread less widely than the horses, however, and never penetrated into western Europe; but they reached Africa as early as three million years ago. In North America, forms like *Camelops* – a large camelid, somewhat intermediate between llamas and true camels – persisted throughout the Pleistocene.

The deer also flourished greatly in Pleistocene times. Their main evolutionary centre lay in Eurasia, but there are secondary centres in

North America (with the *Odocoileus* group of white-tailed and mule deer) and South America, where there is quite a varied assemblage; the earliest deer appeared in this region in the Pliocene. The South American forms were derived from the *Odocoileus* stock.

A central group among Old World deer is formed by the red deer, sambar, barasinga (genus *Cervus*), which are represented by early forms in the Villafranchian; modern species appear in Mosbachian times. The wapiti or American elk is a late invader in North America.

A related group is formed by the giant deer (*Megaceros*), which are

Figure 61. The original centre of evolution of the camel family was North America, where it was represented in Pleistocene times by *Camelops*, a big form measuring some seven feet to the shoulders, and with a mixture of llama-like and camel-like features.

characterised by very large and often very complicated antlers. The record as regards size and span, for deer of all times, is held by the late Pleistocene Irish 'elk', spans of up to 3·5 metres have been cited. Oddly, the lower jaw of these animals was much thickened, so that the jawbone may be practically circular in cross section. The fallow deer (*Dama*) also belongs to this group.

As regards antler complication, the prize goes to the bush-antlered deer (*Eucladoceros*) of the late Villafranchian.

The second main group of Old World deer comprises the mooses (*Alces*), roe deer (*Capreolus*) and reindeer (*Rangifer*). The moose group has a well-documented history from the Villafranchian on; a late Pleistocene offshoot in America is the stag-moose *Cervalces* with its highly lobated antlers. The moose itself entered America at a relatively late date in the Pleistocene, and the same is true for the reindeer (or caribou); its Eurasian history goes back to Mosbachian times.

Figure 62. Most complicated deer antlers of all times may have been those of the bush-antlered deer, *Eucladoceros*, of the Villafranchian in Europe. Span of antlers exactly two metres.

In contrast to the deer, the giraffids were hit very hard by the increasing cold and completely vanished from the northern areas, surviving only in Africa and the Indian region. Here, however, a group of very large giraffe-like animals with antler-like horns was also present (*Sivatherium* and related forms).

The prongbuck family of North America, which had attained its maximum development in the Pliocene, continued to flourish; branched horn-cores are still common in the Pleistocene and in one case a four-horned type appeared (*Tetrameryx*). Twisted horns are also found. Only one species of the family survives to the present day, the pronghorn antelope *Antilocapra americana*.

In contrast, the bovid family, which had its centre in the Old World, reached an unprecedented climax in Pleistocene times. Both in Asia and Africa there were immense troops of the most varied bovid types. In Asia alone some fifty genera are currently recognised for the Pleistocene, and work on the African assemblage is still producing a wealth of new data.

There are records of the highly diverse modern antelope groups, as well as some extinct ones. The present-day African fauna still comprises representatives of most of these groups, while postglacial extinction in

Asia has reduced its antelope fauna. This is true, for instance, for the reduncine antelopes (the kob, waterbuck, reedbuck, etc.) which were distributed in both Africa and Asia in the Pleistocene, but now survive only in Africa. The same holds for the alcelaphines (the gnu and hartebeest group), while the gazelle-blackbuck group or Antelopini, as well as the sable-oryx tribe Hippotragini, persist in both continents.

Most antelopes belong to the temperate or tropical zones, but a few gazelles and especially the saiga group have invaded the higher latitudes.

Figure 63. *Sivatherium*, a great antlered giraffid from the Pleistocene of southern Asia and Africa; shoulder height about seven feet. Females were antlerless.

The saiga, a peculiar-looking animal with oddly swollen muzzle, penetrated both into Europe and Alaska in the Pleistocene (thus being the only true antelope known to have invaded North America) but has since died out in the marginal areas of its range.

The most striking new feature within the Bovidae is the appearance of the true bovines, which evidently evolved from large cattle-like antelopes like the elands, nilgais and wildebeests of the present day. The size tended to increase and the horns tended to become forward-directed – as in fact they are in the wildebeest.

This new type of bovid is first seen in *Leptobos* of the Villafranchian, a lightly-built Eurasian ox, transitional between the antelopes and cattle.

Figure 64. Balearic cave goat *Myotragus*, a specialised insular creature with enlarged, rodent-like gnawing incisors; it stood about 50 cm. at the shoulders. This animal was still in existence in Neolithic times, and its remains are plentiful in the caves of Mallorca.

Various different lineages may be traced back to *Leptobos*-like beginnings: the aurochs group (*Bos*) and its derivative the domestic cattle; the bison group (*Bison*); the Asiatic water buffaloes (*Bubalus*) and so on. The African buffalo group seems to be more distantly related and probably forms an evolutionary line of its own. Here belongs the living caffer buffalo *Syncerus*, as well as the still larger, but more slenderly built *Pelorovis* from Olduvai.

Of all the bovines, *Bison* was the only one to gain access to the New World. Its success here was remarkable, however; in the late Pleistocene, North America was populated by many different kinds of bison, long- and short-horned. The spread of the horns in *Bison latifrons* was more than six feet.

The sheep-goat tribe is also a Pleistocene innovation. The typical sheep and goats are represented in North America by invading forms – the bighorn sheep and mountain goats. The group also contains such forms as the chamois, tahr and goral, all of which had a Eurasian distribution in the Pleistocene. A peculiar form is the Balearic cave goat *Myotragus*, in which the lower middle incisors grew into huge, chisel-like, gnawing teeth like those of the rodents. It was still in existence in Neolithic times.

As we have seen, the Pontian faunas contained a number of musk ox-like animals. True musk oxen (*Ovibos*) appeared in the Pleistocene. Although they originated in the Old World, they entered the New by mid-Pleistocene times and gave rise to some distinct forms like the woodland musk ox.

The faunal exchange between Eurasia and North America was quite intense during the Pleistocene but strictly limited to animals that were equipped to withstand the conditions in Beringia. These ranged from extreme cold (tundra and glaciated areas) to probably cool-temperate at best, with forests of northern type. Among the proboscideans, for instance, the mammoths, but not the straight-tuskers, were able to cross. Among perissodactyls, horses crossed but rhinoceroses did not (which is somewhat surprising as regards the woolly rhino). The artiodactyls that made it across Beringia are typical members of the Arctic or boreal fauna: camels (which are quite hardy Northern animals), reindeer, moose, stag (wapiti), bison, mountain goat and sheep, saiga; while such forms as the prongbucks, giraffids, giant deer, and most antelopes and bovines were unable to cross.

The carnivores are generally quite adaptable and indeed were often successful migrants, which resulted in an increasing similarity between the carnivorous faunas of the two hemispheres. Only the viverrid family remained confined to the Old World, where civets, genets, mongooses and other modern types are found in the Pleistocene. A somewhat similar role was played by the procyonids in the New World – ringtails, raccoons, and coatimundi. In Eurasia this family is represented by the panda group (*Ailurus*). The giant panda (*Ailuropoda*), which is regarded as an ursid by some authors, is particularly common in the Pleistocene of China; several cave and fissure fillings have been found to contain large numbers of remains of these animals, which are now extremely rare.

One of the most remarkable carnivores of all times was probably the hyaenid *Chasmaporthetes* (the closely related Old World *Euryboas* may be identical with it). These animals are found in the Villafranchian of Eurasia, the Villafranchian and mid-Pleistocene of Africa, and the Blancan of North America; this is the only hyaenid known to have invaded the New World. Although clearly a hyena in basic features, it had evolved into an extremely long-limbed, fast-running predator that shows an amazing resemblance to cheetahs both as regards the limbs and even the teeth. To see a hyaenid playing this ecological role seems particularly incongruous. The puzzle is not made simpler by the fact that a large true cheetah (*Acinonyx*) was also present in the Old World Villafranchian, providing strong competition in this field.

Otherwise the hyaenids of the Pleistocene were mainly limited to the modern genera *Hyaena* and *Crocuta*. A giant form of *Hyaena*, closely related to the living brown hyena but the size of a lion, existed in the Villafranchian and middle Pleistocene. In the later Pleistocene of Eurasia and Africa the spotted hyena (*Crocuta*) is the predominant form. This is the cave hyena of Europe, of which there are enormous fossil concentrations in some caves.

The felids of the Pleistocene may be grouped into three main groups. The great majority belong to the modern cat types, including the genera *Panthera* (great cats in the leopard to lion range), *Felis* (pumas, lynxes, smaller cats) and *Acinonyx* (cheetahs). The last-mentioned are restricted to the Old World, but the two other genera are common in both hemispheres. Giant forms of the northern continents are typical of the later Pleistocene – a cave lion in Europe, a giant tiger in northern Asia, and an enormous plains-living cat *Panthera atrox* in North America. Some other forms which are now mainly tropical in distribution, like the leopards and jaguars, ranged far north in Pleistocene times.

A second group of felids is formed by the dagger-toothed cats, which were widely distributed in the Old World in early Pleistocene times. They became extinct there in the middle Pleistocene, but before that succeeded in invading the Americas, where they evolved into the great, sabre-toothed form *Smilodon*. A heavily-built, slow-moving animal, it probably preyed on large, slow-footed game. Remains of *Smilodon* are especially common in the asphalt deposits of Rancho La Brea, Los Angeles.

The third felid group is that of the scimitar-tooth *Homotherium*, which ranged widely in the Old World and North America. In this animal the sabres were shorter, much flattened, and razor sharp; the cheek teeth were extremely thin slicing blades. In Friesenhahn Cave in Texas, skeletons of several adults and cubs have been found, showing that the cave was used as a den. There are also large numbers of immature mammoth bones suggesting that they preyed on juvenile elephants. It is interesting to speculate on the hunting methods of *Homotherium*. No carnivore now living is able to prey on the two present-day elephant species, at least not as a regular source of food; the calves are protected by the elephant herd as a whole. Perhaps the social life of the northern elephants and mammoths was less highly developed.

One of the ancient hyena-like dogs, the Blancan *Borophagus*, survived in the early Pleistocene of North America; but otherwise the canid fauna of the Pleistocene is of modern type. There are wolves, foxes, and raccoon-dogs not unlike the living species. The specialised hunting-dog types, *Lycaon* of Africa and *Cuon* of Eurasia, evolved during the Pleistocene from wolf-like ancestors. The great dire wolves are another offshoot of the wolf group; they are now extinct but formed a dominant element in the tar pit faunas of the West.

North and South America in the Pleistocene were an evolutionary centre

Figure 65. The scimitar-toothed cat *Homotherium*, which was present in the Pleistocene of both hemispheres, had a peculiar, semi-erect stance because the fore limbs were greatly elongated, the hind limbs short and plantigrade. Evidence suggests it may have preyed upon young elephants.

for bears related to the present-day Andean bear (*Tremarctos*). A much larger species inhabited Mexico and the southern USA. Another tremarctine group is that of the gigantic *Arctodus* or short-faced bears. The North American form was quite slenderly built and probably highly predaceous, while the South American arctodonts had heavy, broad teeth possibly suggesting shell-fish feeding habits.

Eurasia, meanwhile, was the homeland of the *Ursus* group of bears. Its largest Pleistocene representative was the European cave bear, but it was closely rivalled by the great brown bears of the Ice Age. The latter now survive in the brown and grizzly bears of the northern hemisphere; the

Figure 66. The great European cave bear *Ursus spelaeus* differs from the related brown bear in its heavier build and, judging from the dentition, more herbivorous habits. Its remains are found in immense numbers in some Pleistocene cave deposits, apparently because the animals died during hibernation.

entrance of this species in America, however, is of fairly late date. The polar bears are a late offshoot of the brown bear group.

The early Pleistocene species of *Ursus* were smaller and more primitive and their characters are essentially conserved in the living black bears; this group entered North America in the middle Pleistocene. The ancient, Pliocene *Agriotherium* survived in the earliest Pleistocene of Europe, but soon became extinct. A recent discovery indicates that agriotheres also invaded South Africa. Otherwise Africa is not known to have harboured any bears except for some Pleistocene populations in the Mediterranean area.

Eurasian mustelids in the Pleistocene are dominated by various ferret-like forms, of which the living marbled polecat is a survivor. Martens, weasels and stoats of various kinds are common in both hemispheres; both the Eurasian and the American type of badger are also present. Striped, spotted, and hog-nosed skunks are found in the American Pleistocene. The wolverine (*Gulo*) evolved in the Old World and invaded America in the middle Pleistocene.

There are some large mustelids, resembling the tayras of South America, in the Villafranchian of Eurasia; they also crossed Beringia and may be ancestral to the living form. The African honey badgers were also distributed in Asia at the time and in fact appear to be of Asiatic origin.

Otters are also common, the genus *Lutra* being widespread in both the Old World and the New. In addition, the shellfish-eating, small-clawed otters (*Aonyx* and related genera), which are now confined to the Old World tropics, were present in Europe. The history of the sea otter (*Enhydra*), too, goes back to the Pleistocene.

Looking at the Pleistocene history of the orders treated above – the Proboscidea, Perissodactyla, Artiodactyla, and Carnivora – the frequency of giant forms is very striking. Some of the mammoths and elephants exceeded all other proboscideans in size. Pleistocene horses were the largest of all times. *Megatapirus* was a giant tapir. The elasmotheriine rhinos were exceeded only by the Tertiary indricotheres. Among the artiodactyls, the deer reach their acme in the moose and giant deer; the bovids in the great Ice Age bisons, buffaloes, and Aurochs; the non-ruminants in the Pleistocene hippopotamus, which was even larger than the living form. There were also giant camels and giant pigs. The European cave lion and North American plains cat were the largest felids of all times. The short-faced hyena was as big as a lion. The short-faced bear *Arctodus* in North America, and the cave bears and brown bears of the Old World, were other giants.

In other orders of mammals the same trend towards gigantism is evident. This even concerns the modest order of the scaly anteaters or pangolins (Pholidota): in the Pleistocene, southeast Asia was inhabited by a giant pangolin more than twice the length of the living Javanese species.

Among the primates, too, the same phenomenon can be seen. One of the largest primates of all times was *Gigantopithecus* of the early and middle Pleistocene in China; the immense jaws of this creature were larger than those of a gorilla. In spite of its great size it has some curiously man-like traits, such as the relatively small incisors and canine teeth; and the cheek teeth are worn down in the same manner as those of primitive men. Perhaps this great ape had a more mixed diet than the other apes, which are mainly herbivorous. A precursor of the Pleistocene *Gigantopithecus* has recently been discovered in the Pliocene of India.

Although no other apes attained such a monstrous size, Pleistocene orangutans were considerably larger than living ones; in the Pleistocene this species was fairly widely distributed in southeastern Asia and ranged north into China. Madagascar was a stronghold of prosimians, some of which also reached great size. The largest of them all, *Megaladapis*, was the size of a two-year-old calf, and several other forms were almost as big.

Some of these animals evidently were still in existence in the eighteenth century.

With the coming of the cold, the apes and monkeys that had inhabited Europe in Pliocene times mostly became extinct. In the Villafranchian two species of monkey are still found there, but only the hardy macaque managed to survive in the later Pleistocene, when it made temporary incursions during the interglacial phases.

Like most other mammalian orders, the Rodentia also produced giants in the Pleistocene. Besides the traditionally large capybaras and related

Figure 67. The Malagasy lemur *Megaladapis*, the largest known prosimian primate, probably still existed in the eighteenth century. Like many other large mammals and birds in Madagascar it probably was exterminated by man.

South American forms (of which one was able to colonise southern North America) may be noted the fabulous North American beaver, *Castoroides*. It was almost as large as a black bear and represents the culmination of a long-range trend towards size increase. The related Eurasian *Trogontherium* was only slightly larger than the living true beavers. Both *Castoroides* and *Trogontherium* belong to an extinct group of beavers which probably were not dam-building.

The rodent fauna of the northern hemisphere was dominated, in the

Pleistocene, by the explosive evolution of the voles and rats; this had got under way in the Pliocene and the proliferation in the Pleistocene was enormous. The vole 'explosion' culminated in the earlier part of the Pleistocene, that of the rats towards the end of the epoch. Among Old World voles the genus *Mimomys*, in which the cheek teeth still have roots, is predominant in the early Pleistocene; later on this group gave rise to the more progressive *Arvicola*, in which the cheek teeth are rootless and grow throughout life. The actual take-over was enacted in the Mosbachian. The northern tundra became the homeland of highly specialised voles like the lemmings and tundra voles. These and other hardy northern forms migrated freely between the western and eastern hemispheres; the marmots entered the Old World, the lemmings the New. Gradually, a circumpolar Arctic fauna came into being.

The Pleistocene was the time of a massive conquest of South America by World Continent mammals. Artiodactyls, perissodactyls, carnivores and rodents multiplied and crowded out many of the local old-timers. But some of the indigenous forms persisted more or less unaffected, especially the Edentata. So far from being overwhelmed by the competition, edentates – especially the ground sloths – pushed successfully into North America and proliferated there as well. Their size culminated in the mighty *Megatherium*; a closely related form is found in southern North America. The related, but much smaller and more lightly-built nothrotheres were common in both continents; in some specimens, mummified skin, muscles and tendons have been found. The family Mylodontidae was mainly deployed in South America, while the Megalonychidae were more frequent in the north. Members of the last-mentioned family also colonised some of the Antillean islands.

The big tank-like glyptodonts, and the related but smaller armadillos, ranged from their South American homeland into southern North America. Both groups were highly varied in the Pleistocene and among the armadillos may be noted the great *Pampatherium* which was up to seven feet long. Pleistocene fossils of anteaters in South America represent both the large ant bear (*Myrmecophaga*) and the smaller tamandua.

Of the other South American forms, the marsupial carnivores were gone, and the litopterns and notoungulates were on the wane; only a few large forms (toxodonts, typotheres and macraucheniids) survived in the Pleistocene. The opossums and other small marsupials, on the other hand, continued to be successful, and the first-mentioned group re-entered North America where opossums had become extinct in the Miocene.

In Australia the native mammals tended to gigantism like everywhere else. *Diprotodon* was the largest marsupial of all times, *Thylacoleo* a great, possibly carnivorous form. Giant kangaroos bigger than any of those now in existence are also found, and the kangaroo-like *Sthenurus* was even larger. There was even a giant form of the lowly duckbilled platypus.

The marine faunas of the Pleistocene are essentially modern, but the cooling of the climate tended to shift Arctic forms like the walrus and the bowhead whale far south of their present-day haunts. Among sirenians, most dugongs definitively vanished from the northern waters; only Steller's sea cow remained in the Arctic, to be exterminated by man in late historical times. The desmostylians had been extinct since the early Pliocene.

The insular faunas of the Pleistocene are in many cases of quite exceptional interest. This is less true of the islands on the continental shelf which thus tended to become joined to the continents when the sea was lowered in glacial times. For instance, the North Sea and English Channel would be dry land and Britain would be part of the European continent. In southeast Asia the islands of Sumatra, Java, Borneo and Formosa would be joined to Asia, while New Guinea and Australia would form a single land mass. In all these islands there would be repeated invasions of land animals from the neighbouring continents, maintaining a fauna essentially similar to the continental one.

But there are other islands that stand in deep water and so have always been isolated by marine straits, even if the regressions led to a narrowing of the gap between the island and the continent. These islands would only be populated by certain types of land animals: either very strong swimmers, or else small animals accidentally rafted on floating trees or the like.

These island faunas show some recurrent evolutionary trends, of which the most striking was the dwarfing of the large mammals. Dwarf elephants, dwarf hippopotami, dwarf deer are found in many islands. The dwarf mammoths of the Santa Barbara islands off California, the dwarf straight-tuskers of the Mediterranean islands, and the dwarf stegodonts and planifrons elephants of the Sunda Islands and Celebes, are good examples; the trend led finally to the evolution of the extraordinary Maltese 'donkey elephant', in which the adult shoulder height was only about three feet.

In the Antilles, the megalonychid ground sloths fared in a similar way. Instead of the great hulking continental beasts, we find a series of ever smaller ground sloths down to extreme forms little larger than a cat.

Probably several factors conspired to bring about these spectacular changes. A size reduction would make the animal more agile in the broken, mountainous terrain typical of most oceanic islands. The limited space and food supply would also make smaller size useful for survival. The lack of dangerous carnivores made it possible to dispense with the protection afforded by large size and strength.

In contrast, small animals like rodents and insectivores often tend to size increase in the insular environment. There are giant rats and dormice in the Mediterranean islands, giant pikas on Corsica and Sardinia, giant lacertilian lizards in the Canary Islands – they all obey the same laws. Perhaps this is also due to the absence of carnivores. As long as cats or foxes may pounce upon you at any time, it may be advantageous to be small so that you can seek shelter in narrow fissures impenetrable to the enemy. But if there are no enemies, you can grow big and fat with impunity. And so it came about that the small animals grew larger, and the large smaller, as if converging on an ideal average size suited for the carefree insular life.

The Pleistocene fauna of Madgascar would deserve a chapter of its own. The giant lemurs have already been mentioned, but there were many other remarkable creatures as well. There was a miniature hippopotamus, a large form of the fossa (a cat-like viverrid carnivore), and the big elephant bird *Aepyornis*, whose eggs held up to eight litres and are the largest known eggs. Madagascar is a minor continent rather than an island and so the special insular trends were less important here, except as regards the Malagasy hippo.

Some of the insular faunas have survived up to historical times: this is certainly true for the moas of New Zealand, the giant lemurs and elephant birds of Madagascar, and the giant pikas of Sardinia. Only a few thousand years separate us from the time of the glaciers and giant animals of the Pleistocene. In geological perspective, the faunal revolution at the end of the Ice Age appears instantaneous – a catastrophe of incredible swiftness. And with it, the Age of Mammals may well be said to come to an end. We enter a new phase in the history of the earth: the Age of Man.

Chapter 10

The age of man

Man: king of ashes.
Harry Martinsson: *Aniara.*

REMAINS OF hominids in Tertiary deposits are extremely rare and their status is still disputed by many authorities. When we approach the Pleistocene, however, fossil remains of beings generally accepted as Hominidae begin to occur. The earliest dated so far come from Omo in southern Ethiopia by Lake Rudolf and are almost four million years old, which places them in the latest Pliocene. All the other finds are more recent than three million years, the beginning of the Pleistocene.

Even in the Pleistocene hominid fossils are quite rare and suffice only to give a rather sketchy picture of the evolution of man within the last two million years or so. To be sure, if we include other traces of human activity – in particular the stone artifacts fashioned by early man – hominid remains are plentiful in Pleistocene deposits in some areas. Unfortunately it has turned out that the artifact assemblages do not give any reliable information on the anatomical characters of their authors. If we want to know anything about the physical presence of early man, we have to turn to the meagre harvest of fossil bones and teeth.

From the late Pliocene and Villafranchian – between roughly four and one million years ago – we know of a number of hominid fossils from Africa. No hominid remains of this age are known in Europe; in Asia there may be a few but their age is still disputed.

Most or all of the African finds from the Villafranchian may be grouped broadly in the genus *Australopithecus*. The name, meaning 'southern ape', is somewhat unhappily chosen as these beings were very far from apes; however, the rules of zoological nomenclature do not permit any change of name. Specimens of *Australopithecus* have been found in old cave fillings in the Sterkfontein Valley and Makapansgat, Transvaal; at Taungs, Botswana; at Omo, Ethiopia; at Olduvai Gorge, Tanzania, as well as a few other sites. Most of the finds are evidently Villafranchian in date but the

Ethiopian are slightly older and a number of finds are more recent; they fall within the Swartkrans faunal stage referred to the early part of the middle Pleistocene, approximately equivalent to the Mosbachian of Europe.

Outside Africa, some hominid remains, for instance from China and Java, have been referred to *Australopithecus*. If this is correct, the geographic range of these hominids would seem to have been roughly the same as that of the late Tertiary *Ramapithecus*, with the exception of Europe.

With an average stature of four feet, *Australopithecus* was about the same height as a modern pygmy, and had a fully upright posture. This is proved by such anatomical characters as the position of the neck joint beneath the skull (which balanced on a vertical neck), and the shape of the vertebrae, hip bones, legs and feet.

Some anatomical traits however suggest that his adaptation to bipedality may have been less advanced than that of ourselves. The important gluteal musculature (which fashions the typical human buttock) was evidently less developed in *Australopithecus*, and his gait may have been waddling rather than striding, although he may have been quite a good runner. Another primitive feature is the rather large size of the middle toe, while the big toe is smaller than in a modern human foot.

In the hands, the thumb appears to have been slightly shorter in relation to the other fingers than in ourselves. It is just possible that this is a legacy from arboreal life. In swinging hand over hand, a reduction of the thumb and lengthening of the other fingers is advantageous because it makes it easier for the fingers to hook onto a branch. Indeed the living apes have very long fingers and short thumbs. This, on the other hand, makes the hand less adapted for seizing and manipulating – actually, apes may prefer to use their feet rather than their hands to do this. But however this may be, the limbs of *Australopithecus* are essentially human.

This is also true for the dentition, even if there are some differences in relative proportions: the cheek teeth are larger than in modern man, the front teeth smaller. As we have seen, a hominid type of dentition may be traced back to the Miocene and perhaps even to the Oligocene, so that this should not be surprising. What *Australopithecus* does teach us is that early man had attained a fully upright posture by early Pleistocene time.

Modern man, however, differs from other primates not only in dentition and posture but also in possessing a much larger brain. In this respect, *Australopithecus* was still quite primitive. The average braincase volume was about 450 c.c., which is nearly the same as in the gorilla, and one-third of that in modern man (about 1,350 c.c.).

In spite of his shortcomings as regards brain volume (and, perhaps, manual dexterity) *Australopithecus* fashioned various kinds of stone and bone tools; it is only at the oldest site (Omo) that artefacts are definitely missing. His so-called pebble culture, of which the Oldowan is a good example, features pebbles from which one or a few flakes have been chipped off to produce a working edge – very likely used to cut up the game animals that he killed for food. At Olduvai he also erected small circular stone shelters. At Makapan, a number of baboon skulls have been found with lesions suggesting that *Australopithecus* used bone clubs to kill his prey; the direction of the blow indicates that the proportion of right-handed to left-handed was 95:5, much as in modern man.

Two forms of *Australopithecus* are generally recognised – most probably two species. They differ in the dentition and skull architecture. *Australopithecus africanus* is the most man-like form, with comparatively small cheek teeth (though still bigger than in modern man). In the other form, *A. robustus*, the cheek teeth are much larger and the front teeth still smaller; the jaws are enormous and the jaw musculature must have been huge, because the muscles met at the crown of the head, where a crest developed for their insertion – a unique character in hominids as far as we know.

The difference in the jaw apparatus suggests that the two forms had different diets. *Australopithecus africanus* seems to have hunted for animal food, while the *robustus* type may have been more vegetarian, perhaps eating tubers and the like; the grit introduced with the food would cause heavy wear of the cheek teeth. Perhaps in connection with the stresses caused by the powerful muscles, *robustus* also had much larger eyebrow ridges than *africanus*; also his head was much flatter, while *africanus* has a distinctly rounded head.

In the Sterkfontein, Makapan and Taungs deposits, the smaller type, *A. africanus*, occurs in the early (presumably Villafranchian) fauna, while the big *A. robustus* is present in the mid-Pleistocene Swartkrans stage. The geological evidence shows that the climate of the Sterkfontein valley was rather dry in the time of *A. africanus*, suggesting a steppe or savanna environment. Later on, the climate became much more humid, and the *robustus* men of the middle Pleistocene probably lived in a forest milieu in the Sterkfontein valley.

At Olduvai there is no such separation between the two forms. Both have been found at early levels dated at about 1·75 million years. In the middle Pleistocene, however, only *A. robustus* survived. It is thought that *A. africanus* had vanished at that time because it had already given rise to more advanced forms of man, present in the middle Pleistocene both at Olduvai and Swartkrans.

The Olduvai form of early *Australopithecus* is in fact so man-like that some authorities prefer to regard it as a member of the genus *Homo* and call it *Homo habilis*. Other students would consider it a local variant belonging to the species *A. africanus*.

Going on to the middle Pleistocene, we can note the appearance of a new type of man replacing the ancient *Australopithecus*. The new type used to be called *Pithecanthropus* but is nowadays generally regarded as a species of our own genus *Homo*: *H. erectus*. Finds of this species are widespread (China, Java, North Africa, Olduvai, Swartkrans – at the two last-mentioned sites in association with *Australopithecus robustus*). In early forms of *H. erectus*, probably dating from 0·6 to one million years back, the average braincase volume is still rather small, only about 850 c.c. In later populations the average increased up to some 1,050 c.c.

With an average height of five feet, *Homo erectus* had grown a little taller than his ancestor and clearly was more man-like, although his flattened skull with the great eyebrow ridges and the deep constriction behind the eye-sockets is still very primitive looking. Cultures attributed to this kind of man show important advances beyond the *Australopithecus* stage and the use of fire.

At the same time (the Mosbachian), Europe was inhabited by men resembling *Homo erectus* in some characters but endowed with much larger brains (over 1,400 c.c. in the one known specimen, part of a skull from the Elsterian glaciation, found at Vértesszöllös in Hungary). A mandible from Mauer near Heidelberg, dating from the Cromerian interglacial, probably belongs to the same type of man, and also differs in some important respects from typical *H. erectus*.

Unfortunately, knowledge of the Mosbachian humans in Europe is very fragmentary at present. The material from the later middle Pleistocene is somewhat better, consisting of parts of two skulls from Swanscombe in England and Steinheim in Germany (both from the Holsteinian interglacial, *ca.* 250,000 years) and a jaw from the cave of Montmaurin, France (probably from the Saalian glaciation). The jawbone is smaller and more lightly built than the Heidelberg one; the skulls are well rounded almost as in modern man, and the braincase capacities range from 1,150 to 1,300 c.c. The eyebrow ridges, however, are still well developed. It has been suggested that both modern man and Neandertal man, as diverging evolutionary lines, may be derived from this type of early man.

At the same time, and even much later, parts of Asia and Africa were still inhabited by men of essentially *Homo erectus* type, with brain capacities averaging little more than 1,100 c.c. The most recent of these may be Broken Hill man of Zambia, who has been tentatively dated at 30,000 years. This is a time when even the true Neandertaler of Europe had already vanished and men of modern type were settling in various parts of the Old World.

Before that, the Neandertal men of the Eemian interglacial and the earlier part of the Weichselian glaciation had held sway for many thousands of years. Retaining such primitive features as the big eyebrow ridges, flattened skull and sloping chin, they were highly advanced as regards cranial capacity: the average value is in fact higher than in modern man. Short and stockily built, the Neandertalers seem to have been unusually powerful, with barrel-chested bodies, bull necks, and peculiarly curved thighbones.

Burials are not uncommon, either of the entire body or just the head; there are also mementoes of other ritualistic behaviour. Clearly, Neandertal man pondered over the mysteries of life and death. Outside Europe, men of Neandertal type are known in North Africa, the Levant, and some other parts of Asia. (See Plate 11.)

Meanwhile, early types of modern *Homo sapiens* make their appearance. A very early find of essentially modern type comes from a cave at Fontéchevade in France and evidently dates from the Eemian interglacial. These skulls lack completely the heavy eyebrow ridges of the Neandertalers and have a typical *Homo sapiens* brow. Their relationships to Eemian Neandertalers (who were liberally endowed with eyebrow ridges) are puzzling.

In Africa, some skulls from Kanjera, Kenya, are quite modern in type but may be almost as ancient as Fontéchevade man. At Omo, too, ancient *Homo sapiens* have been found. A cave in Borneo has yielded a modern-type skull which has been radiocarbon dated at about 40,000 years. Other modern *Homo sapiens* from the Levant and Europe may be dated at 30–35,000 years, suggesting perhaps an invasion from the southeast.

Soon after the entrance of modern man in Europe, the material culture underwent important change and evolution. The flint industry of these Cro-Magnon men was very advanced, and the marvels of cave art were produced.

Modern *Homo sapiens* spread into all corners of the inhabitable world. Early tribes perhaps related to the bushmen appeared in South Africa; Mongoloid peoples colonised eastern Asia. As early as 30,000 years ago, Australia was inhabited by man – probably ancestral to the living Australoids.

The exact date of human entrance in North America is still disputed. There may have been an early immigration more than 25,000 years ago, before the vast Wisconsinian ice-field formed an imprenetrable barrier from the Pacific coast to the Atlantic. If so, we have only meagre and uncertain information of their presence, and none at all as regards their physical character. After the bipartition of the ice sheet about 13,000 years ago, Mongoloid tribes migrated southward through the ice-free corridor east of the Rocky Mountains and established themselves in the plains south of the ice margin; these Paleo-Indians specialised in big game hunting. Two thousand years later, men had reached the southern tip of South America.

Always the presence of advanced human hunting cultures seems to be connected with a sudden decline in the large game. The correlation is so close that it is hard to escape the conclusion that the extinction was caused in some way by the influence of man.

At present it seems that the earliest episode of large-scale extinction occurred in Africa at some time roughly 60,000 years ago. (The same may be true of tropical Asia but this has not yet been studied in similar detail.) The men of this time were, culturally, late Acheulians – that is to say, they had a highly developed hunting culture of the hand-axe tradition.

They also used fire, and it has been thought that repeated firing was the main agency by which the animal world was depleted. There is evidence of forest firing in Europe as early as in the Holsteinian interglacial, and the deforestation of much of America and Australia may be due to the 'peripatetic pyromania' of aboriginal groups. By burning off the vegetation of wide areas, probably just in order to drive game, early man destroyed the environment and so wrought an immensely greater havoc than that caused by hunting alone.

However this may be, almost forty per cent of the genera of larger mammals in Africa became extinct. These include giant baboons, scimitar-toothed cats, three-toed horses, a great number of different kinds of Suidae, the African giant deer, the peculiar antlered giraffe *Libytherium* (which resembled the Indian *Sivatherium*), and various forms of buffalos and antelopes. The remaining, impoverished fauna of large mammals (or 'megafauna') is essentially the one that is still in existence in Africa.

In Europe a somewhat similar story was enacted but at a later date – roughly during the interval 30–10,000 years ago. This, it should also be noted, is the time of the *Homo sapiens* hunting cultures in Europe. The animals that became extinct during this interval include the straight-tusker and the mammoth; the woolly rhino and Merck's rhino; the Irish 'elk', musk ox and steppe bison; the cave hyena, cave lion, cave panther, cave bear and scimitar cat. The great majority of the animals that became extinct in Europe belonged to the megafauna; only about fifty per cent of the species of the larger mammals remained alive.

The process of extinction in Europe, and probably also in Africa, seems to have been quite protracted; 20,000 years or more elapsed, during which time one species after the other became rarer and finally vanished. In North America, on the other hand, extinction was swift and catastrophic and may have occurred within little more than a thousand years.

Animals that vanished in this episode of mass death include at least three species of ground sloth, the giant beaver *Castoroides*, the great plains cat *Panthera atrox*, two species of peccary, one form of antelope, and probably many others. A number of forms may have held out a little longer, including the indigenous American camels (two genera), horse, dire wolf, and American mastodon. At best, however, they survived for another millennium or two, and it is clear that a total of about seventy per cent of the North American megafauna died out rapidly at the end of the Wisconsinian glaciation, immediately after the immigration of hunting Paleo-Indian tribes.

The swiftness of the catastrophe, compared with the processes in the Old World, may well be due to the fact that the indigenous American mammals had no previous experience of man and so may have lacked the flight responses that would have been essential for survival. It is noteworthy that most of the Eurasian invaders in North America – the moose, wapiti, caribou, musk ox, grizzly bears, and so on – were able to maintain themselves, perhaps because of their long previous conditioning to man.

In South America the extinction seems to have been equally rapid. Early human cultures here are contemporary with the last records of many large mammals such as ground sloths and indigenous horses.

In Australia, too, the entrance of man was followed by a large-scale reduction of big marsupials. Over thirty per cent of the larger species became extinct, though the exact time-table of these events cannot as yet be determined.

When we come to insular faunas, the juxtaposition of human invasion and destruction of the local fauna is so clear that only one solution is possible. Madagascar, for instance, was colonised by man at a relatively late date – towards the end of the first millennium AD. Less than a thousand years later the remarkable local lemur species had been reduced by more than forty per cent. All of the forms that became extinct were fairly large, and nearly all were probably diurnal in habits, while those that survive are mostly nocturnal (the size of the eye-sockets gives a good clue to habits). Naturally, animals that are about in the daytime are easier to kill than nocturnal ones. With the giant lemurs went also the elephant birds, which up to then seem to have nested in large colonies in the coastal plains of Madagascar. Egg-robbing may have played an important part in their extermination.

An equally clear-cut case is the destruction of the moa birds of New Zealand after the invasion of human tribes about AD 800. Before AD 1600 all of the twenty-two species of moa known to have co-existed with man in New Zealand had become extinct.

It has also been suggested that extinction in the late Pleistocene might be due to a rapid change of climate. But the extinction in different parts of the world is not synchronous and does not always coincide with any definite climatic change. On the other hand it does seem always to coincide with the appearance of advanced human hunting cultures. So the evidence seems definitely to point to man's interference as the basic factor, perhaps due mainly to his influence on the environment and the megafauna by burning rather than by actual overkill. Even so, the devastation by direct hunting apparently also played an important part, especially in the destruction of insular faunas. In more recent years, the dodo in Mauritius and the great auk of the North Atlantic skerries were certainly slaughtered to extinction.

The large animals that survived the mass extermination in many cases show a remarkable decrease in size since Pleistocene times. Examples may be taken at random from almost any part of the world. In Europe the brown bear, the pardel lynx, the red deer, the moose and the wisent are all smaller than their Pleistocene forbears. The aurochs of the postglacial was smaller than that of the late Pleistocene. The living polar bear is smaller than that of the Weichselian. In North America, size reduction has occurred in many forms, although studies at present are incomplete; a spectacular case is the black bear, for late Pleistocene blacks are as big as modern grizzlies.

Dwarfing since Pleistocene times may also be noted in the orangutan of southeast Asia, the tigers of Asia and lions of Africa, the tapirs of Malaya, the rhinoceroses of Sumatra and Java, the fossa of Madagascar, and numerous other instances. And yet dwarfing is not universal. In Europe, for instance, where some large mammals were reduced in size, others – like the Northern lynx and the wolf – show little or no size reduction. The great brown bears of Alaska and British Columbia, as well as the grizzly bears, are the same size as their late Pleistocene progenitors.

We know some of the laws or rules governing size changes in mammals. In the long perspective, there seems to be a general tendency to size increase, because this will make you a little stronger, may give you a slightly longer life, and make your metabolic balance slightly more efficient; consequently, most mammalian evolutionary sequences show gradual phyletic growth. A good example is the horse lineage from tiny eohippus (*Hyracotherium*) of the Eocene to the large modern *Equus*. On the other hand, size increase is generally checked by the special demands set by the current mode of life of the animal. It would not do for a panther to grow to the size of a lion, because the lion is already there to do the predation suitable for an animal of that size.

Reduction in size is less common, but it does occur in some instances, and it can often be shown to be advantageous under the circumstances. In a warm climate, for instance, smaller size will lead to increased heat-loss. The smaller you become, the greater the ratio between surface and bulk, and the faster you get rid of excess heat. Pygmies are typical products of the tropical climate.

The same factor may also be seen at play in some glacial-interglacial lineages during the Pleistocene. In Europe, for instance, the brown bear was regularly reduced in size in interglacial times, and again grew larger as the climate became colder. Perhaps some of the cases of postglacial dwarfing are simply due to the change in climate.

In some mammals the relationship between climate and size operates the other way round – these forms become smaller as the climate gets colder. The ermine is a good example; Scottish stoats are smaller than English ones, Scandinavian stoats smaller than central European ones. The same is true for the weasel; in fact the weasels of southern Europe are almost as big as the Scandinavian ermines. It is not surprising, then, to learn that interglacial stoats tended to be larger than those of the glacial intervals.

Still another factor leading to smaller size is lack of space or food, as in the case of insular animals. By becoming smaller, the animals reduce their food intake and territorial demands. The result will be the same if the environment deteriorates, for instance through the agency of man. Fires, and the introduction of domestic animals, will reduce the space and food available to wild animals. If the deterioration is slow enough, the animals may be able to react by suitable adaptive changes in size; if it is too rapid, the hardest-hit species may become extinct. Thus it would seem that extinction and reduction in size may have had a common denominator in many instances.

In many parts of the world it seems indeed that this is the only possible explanation. In southeast Asia, for instance, the climate probably remained much the same at the end of the last glaciation, so that there can hardly be any climatic explanation of the size reduction in the tapirs, rhinoceroses, tigers and orangutans. Man, however, has been present in this area during the whole of this crucial time and has presumably been waging his usual struggle against nature for many thousands of years.

Analysing the size changes in mammals in relation to the evolving human cultures can be particularly instructive. Towards the end of the last glaciation, for instance, various carnivore species inhabiting the eastern coasts of the Mediterranean tended to become smaller. This is true for the bear, the (spotted) hyena, the wolf, the wild cat and the beech marten. The trend was rather slow at first, but at the transition to the so-called Mesolithic cultural stage the dwarfing became very rapid, as if the environment had suddenly become adverse. The hyena, indeed, became almost ridiculously small and then died out completely. (The hyena now present in this area is another, smaller species, the striped hyena.)

One of the carnivores, however, did not obey this law of general dwarfing. The red fox, in fact, grew bigger than ever before in Mesolithic times, as if it thrived in the circumstances so adverse to most of the carnivores. Perhaps it did well in some scavenging role around human settlements. Later on, with the introduction of cultivation in the Neolithic, the fox diminished in size.

Increasingly, and at an accelerating rate, man has been putting his stamp on his environment during the brief span of the postglacial – or, as we should perhaps call it, the Flandrian interglacial. Forest firing and cultivation, animal husbandry with consequent depletion of wild game, the fight against the predators that man persists in regarding as enemies and competitors – all this has been going on for a long, long time. Now come other, much faster processes: the conversion of land into built-up areas, the spread of gigantic highways, the pollution of air, water, and living organisms.

The Age of Mammals is over and we are in the Age of Man, for better or worse. Will it be a long or a short age? We do not know; but if man persists in fighting his futile war against nature, the chances are it will be short – perhaps no more than the brief Flandrian interglacial.

For, while we fondly believe ourselves to be the lords of creation and the masters of nature, all of the factors of evolution are still at work within and around us. Natural selection is still there. You sit behind the steering wheel and little dream that you, too, are under its influence: your temper, the acuteness of your senses, the adaptiveness of your reactions, are continuously being tested.

Of course we do affect natural selection, though rarely in a sensible way. Every time we introduce a new substance like a stimulant or drug into the complex fabric of our bodies, selection will patiently start on the long and perhaps hopeless road towards immunity. It will kill us right and left, by lung cancer, thrombosis and the like; yet perhaps just too late in life to affect reproduction and so give generations a thousand years hence immunity to tobacco, or pot, or whatever new poison man has decided to take up. We introduce the same kind of selection when we kill bugs with insecticides, but here the resistance comes much faster because their generations are so much shorter.

On top of everything else, man is facing serious over-population, which so far as we know has no precedent in the Age of Mammals. Unless this problem is solved in the near future, the imbalance between man and nature will continue to grow and our environment may become damaged beyond repair.

What if man should become extinct? At present, the fauna of the earth gives little hope of any other being evolving the same kind of intelligence, involving true speech and conceptual thought. This is true even though recent studies of dolphins and other cetaceans are giving fascinating insight into other types of high intelligence. Yet these animals are too specialised, too locked in their own world to become dominant creatures in the same way as man. The Age of Dolphins is a highly improbable concept.

If *Ramapithecus* had become extinct, it is just possible that *Oreopithecus* might have evolved into a man-like being. But at present there does not seem to be any primate that could give rise to anything resembling the human species. Certainly the true apes are excluded from any such possibility by their own specialisation for other ways of life.

There are also less gloomy views of the future of the Age of Man. In the opinion of Julian Huxley, man is the only animal which now has an evolutionary future. All the other animals are in evolutionary blind alleys and have already realised all their potentialities. Their specialisations make any radical departure impossible. The horse is a horse and can become nothing else, except perhaps just a little horsier. The rat is a finished creature and the only thing it can give rise to is perhaps a slightly better and more efficient rat. The other mammals, too, have run their course. We have seen that the Miocene and Pliocene were epochs of gradual perfection in the mammalian species, and now there is little left to be done. In a way, perhaps, it seems logical that the Age of Mammals should come to an end at this stage.

Man, on the other hand, has all his evolutionary potential still in store. He is unspecialised in most characters; he is specialised in being a perfect biped, and in the possession of brain power involving speech and conceptual thought. This is a very recent innovation in evolutionary history, and – in Huxley's opinion – it is the only one that can now lead to any major biological progress.

In this vision there are glittering prizes that may be within the reach of our descendants. Perhaps, in future, man will populate other worlds besides the earth that bore him; perhaps he will reach out towards the ends of the universe. That we cannot say now. But looking back on the long panorama of Cenozoic life, I think we ought to sense the richness and beauty of life that is possible on this earth of ours, and realise that equal splendour may again be attained in the future, in new guises, under the aegis of wise and benevolent men. And we should also realise that man, who is now planning for his own or the next generation, must begin to plan also for the geological time that is ahead of him. It may stretch as far into the future as time behind us extends into the past.

Appendix

A LIST of the tetrapod orders in existence during the Cenozoic Era.

Class Amphibia: tetrapods showing metamorphosis during development.
Order Anura: frogs and toads, Jurassic to Recent.
Order Urodela: salamanders and newts, Jurassic to Recent.
Order Apoda: tropical limbless amphibians, Recent.
Class Reptilia: tetrapods in which development is direct.
Order Chelonia: turtles, Triassic to Recent.
Order Squamata: lizards and snakes, Triassic to Recent.
Order Rhynchocephalia: tuataras and their allies, Triassic to Recent.
Order Choristodera: champsosaurs, Cretaceous to Eocene.
Order Crocodilia: crocodilians, Triassic to Recent.
Class Aves: feathered tetrapods.
Order Tinamiformes: tinamous, Pleistocene to Recent.
Order Struthioniformes: ostriches, Eocene to Recent.
Order Rheiformes: rheas, Miocene to Recent.
Order Casuariformes: emus and cassowaries, Pliocene to Recent.
Order Aepyornithiformes: elephant birds, Eocene to Subrecent.
Order Dinornithiformes: moas, Miocene to Subrecent.
Order Apterygiformes: kiwis, Pleistocene to Recent.
Order Gaviiformes: loons, Cretaceous to Recent.
Order Podicipediformes: grebes, Cretaceous to Recent.
Order Procellariiformes: albatrosses, shearwaters and petrels, Eocene to Recent.
Order Sphenisciformes: penguins, Eocene to Recent.
Order Pelecaniformes: pelicans, cormorants, etc., Cretaceous to Recent.
Order Ciconiiformes: wading birds, Cretaceous to Recent.
Order Anseriformes: ducks, geese and swans, Eocene to Recent.
Order Falconiformes: raptorial birds, Eocene to Recent.
Order Galliformes: fowl and related birds, Eocene to Recent.
Order Ralliformes: cranes and rails, Eocene to Recent.
Order Diatrymiformes: terror cranes, Paleocene to Eocene.
Order Charadriiformes: gulls, auks, plovers, etc., Cretaceous to Recent.
Order Columbiformes: doves and pigeons, Miocene to Recent.
Order Psittaciformes: parrots, Miocene to Recent.
Order Cuculiformes: cuckoos, Eocene to Recent.
Order Strigiformes: owls, Eocene to Recent.
Order Caprimulgiformes: nightjars, Pleistocene to Recent.
Order Apodiformes: swifts and hummingbirds, Eocene to Recent.
Order Coliiformes: colies, Recent.
Order Trogoniformes: trogons, Eocene to Recent.

Order Coraciiformes: kingfishers etc., Eocene to Recent.
Order Piciformes: woodpeckers, Miocene to Recent.
Order Passeriformes: perching birds, Eocene to Recent.
Class Mammalia: tetrapods with hair, that suckle their young.
Subclass Prototheria: egg-laying mammals.
Order Monotremata: monotremes, Pleistocene to Recent.
Subclass Allotheria: early, specialised mammals.
Order Multituberculata: multituberculates, Jurassic to Eocene.
Subclass Theria: higher mammals.
Infraclass Metatheria: marsupials.
Order Marsupicarnivora: opossums, marsupial carnivores, etc., Cretaceous to Recent.
Order Paucituberculata: opossum-rats, etc., Paleocene to Recent.
Order Peramelina: bandicoots, Pleistocene to Recent.
Order Diprotodonta: phalangers, kangaroos, etc., Oligocene to Recent.
Infraclass Eutheria: placental mammals.
Order Insectivora: insectivores, Cretaceous to Recent.
Order Tillodontia: tillodonts, Paleocene to Eocene.
Order Taeniodontia: taeniodonts, Paleocene to Eocene.
Order Chiroptera: bats, Eocene to Recent.
Order Primates: primates, Cretaceous to Recent.
Order Creodonta: creodonts, Cretaceous to Recent.
Order Carnivora: carnivores, Paleocene to Recent.
Order Condylarthra: condylarths, Cretaceous to Miocene.
Order Pantodonta: pantodonts, Paleocene to Oligocene.
Order Dinocerata: uintatheres, Paleocene to Eocene.
Order Xenungulata: xenungulates, Paleocene.
Order Pyrotheria: pyrotheres, Eocene to Oligocene.
Order Proboscidea: mastodonts and elephants, Eocene to Recent.
Order Sirenia: sea-cows, Eocene to Recent.
Order Desmostylia: desmostylians, Miocene to Pliocene.
Order Embrithopoda: arsinoitheres, Oligocene.
Order Hyracoidea: hyraxes, Oligocence to Recent.
Order Notoungulata: notoungulates, Paleocene to Pleistocene.
Order Astrapotheria: astrapotheres, Paleocene to Miocene.
Order Litopterna: litopterns, Paleocene to Pleistocene.
Order Perissodactyla: odd-toed ungulates, Paleocene to Recent.
Order Artiodactyla: even-toed ungulates, Eocene to Recent.
Order Edentata: edentates, Paleocene to Recent.
Order Pholidota: scaly anteaters, Oligocene to Recent.
Order Tubulidentata: aardvarks, Eocene to Recent.
Order Cetacea: whales, Eocene to Recent.
Order Rodentia: rodents, Paleocene to Recent.
Order Lagomorpha: hares and rabbits, Eocene to Recent.

N.B. Only the life span actually demonstrated by fossil finds has been indicated. In many cases the term of existence must have been much longer; for instance, the Monotremata undoubtedly arose in the Mesozoic, although no pre-Pleistocene fossils have so far been found.

Notes

A GENERAL survey of vertebrate history may be found in Romer (1967); see also Colbert (1969), Kuhn-Schnyder (1953) and Romer (1968). Other paleontological texts are Abel (1912, 1927), Simpson (1953), Zittel (1925). Paleobotany: Arnold (1947), Mägdefrau (1968); the last-mentioned also contains excellent restorations of ancient scenes. The Cenozoic era and its fauna or flora are treated in Osborn (1910), Papp & Thenius (1959), Pearson (1964), Scott (1962), Thenius (1959), Thenius & Hofer (1960). Detailed climatic history in Nairn (1961). On paleogeography see Schuchert (1955) and Wills (1952), on mammalian classification Simpson (1945). Texts in historical geology: Brinkmann (1967), Bubnoff (1956), Dunbar (1949), Kummel (1961), Woodford (1965).

Chapter 1 Fossils and the origin of mammals

On strontium in fossil bones see Toots and Voorhies (1965). The Age of Reptiles has been treated e.g. by Colbert (1965), Kurtén (1968 a), Swinton (1962). The subdivision of the Tertiary period is due to Lyell (1875). On the dating of the Tertiary by the potassium-argon method see Evernden & James (1964), Evernden *et al.* (1964). Hildebrand (1958) compares the evolutionary process with cosmic events. The paleotemperature method has been used particularly by Emiliani (1958, 1961). On paleomagnetism see Irving (1964), Runcorn (1962). Numerous studies on sea-floor spreading and continental drift may be found in *Nature* and *Science*, especially from 1967 on. Good recent surveys are Dietz *&* Holden (1970) and Menard (1969). The transition from reptile to mammal is discussed e.g. by Brink (1956), Kermack (1965), Simpson (1962). On the effects of the splitting up of continents in the early Tertiary see Kurtén (1969). Sloan & Van Valen (1965) deal with the transition from the Cretaceous to the Tertiary in Montana.

Chapter 2 The Paleocene: epoch of conquest

Paleocene mammals of North America are treated e.g. by Gazin (1956), Jepsen (1940), Matthew (1937), Simpson (1937), those of Europe by Russell (1964); a European life scene is described by Russell (1962). Kurtén (1966) discusses land connections between the Old World and the New in the early Tertiary. Matthew & Granger (1925 a) describe the fauna of Gashato; new studies of early Mongolian mammals are at present being made by Mongolian and Polish students.

Chapter 3 The Eocene: epoch of consolidation

Eocene mammals of North America are treated *i.a.* by Gazin (1962), McKenna (1960), Matthew & Granger (1915–1918), those of Europe by Depéret (1917), Stehlin (1903–1916), Teilhard (1922). On the earliest bat see Jepsen (1966), on Paleocene eohippus Jepsen & Woodburne (1969). A vivid picture of the Geiseltal is given by Krumbiegel (1959). The fauna of Mongolia *i.a.* in Matthew & Granger (1925 b), that of southern Asia in Colbert (1938) and Dehm & Oettingen-Spielberg (1958). R. J. G. Savage is preparing a study of Africa's early Tertiary fauna. On marine mammals (whales) see Kellogg (1936).

Chapter 4 The Oligocene: epoch of transition

The classic on North American Oligocene mammals is Scott, Jepsen & Wood (1941). Important early works in Europe are Filhol (1882) and Filhol (1876–1877), the latter on the Phosphorites of Quercy. See also Lavocat (1951). Asiatic finds have been described e.g. by Bohlin (1942–1946), Gromova (1954), and African by Schlosser (1911). The primates of Fayum were presented by Simons (1967).

Chapter 5 The Miocene: epoch of revolutions

Mould in basalt of a rhinoceros: Chappell *et al.* (1951). Miocene mammals of North America have been described e.g. by Downs (1956), Matthew (1924) and Wilson (1960); those of Europe by Crusafont, Truyols & Villalta (1955), Thenius (1952) and others. Bishop (1967) summarises African faunas. Asiatic mammals are described by Colbert (1935), marine mammals by Mitchell (1966). On paleotemperatures see Emiliani (1961).

Chapter 6 The Pliocene: epoch of climax

A classic work is Gaudry (1867). A long series of papers in *Palaeontologia Sinica* by Bohlin, Zdansky and others describe the Chinese *Hipparion* fauna. On southern Asia see Colbert (1935). *Ramapithecus* in Pilbeam (1966). For Pliocene mammals in North America see e.g. Savage (1951). Pliocene climatic oscillations are described by Zagwijn (1967).

Chapter 7 Australia

On fossil mammals see Ride (1964), Woodburne (1967) and others. The moas of New Zealand are treated by Kröschke (1963), the zoogeography of the Australian region by Simpson (1961).

Chapter 8 South America

Ameghino's *Obras Completas* contain the life work of the great pioneer. Later studies include, *i.a.*, Loomis (1914), Paula Couto (1952), Scott (1912), Simpson (1967).

Chapter 9 The Pleistocene Ice Age

Survey of the Ice Age in Kurtén (in the press). See also Butzer (1964), Charlesworth (1957), Flint (1957), Woldstedt (1969), Zeuner (1959, 1961). On paleotemperatures see Emiliani (1966). The role of Beringia is treated by Hopkins (1967), the Pleistocene of North America in Wright & Frey (1965). On European Pleistocene mammals see Kurtén (1968 b), Toepfer (1963). Asiatic mammals e.g. in Pei (1934), African in Leakey (1965), Cooke (1967); on Rancho La Brea see Stock (1930).

Chapter 10 The Age of Man

Human evolution in Howell (1969), Howells (1967), Le Gros Clark (1964, 1967), Oakley (1964) and many others. The extinction is treated in Martin & Wright (1967); see also Merrilees (1968). Size changes in the mammals of the Levant are discussed by Kurtén (1965). Man's evolutionary future has been treated e.g. by Huxley (1942) and Simpson (1960).

References

Abel, O. (1912) *Grundzüge der Paläobiologie der Wirbeltiere.* Stuttgart.
(1927) *Lebensbilder aus der Tierwelt der Vorzeit.* Jena.

Arnold, C. A. (1947) *An introduction to paleobotany.* New York.

Bishop, W. W. (1967) 'The later Tertiary in East Africa – volcanics, sediments, and faunal inventory.' In: Bishop & Clark (ed.): *Background to evolution in Africa,* 31–56. Chicago.

Bohlin, B. (1942–46) 'The fossil mammals from the Tertiary deposits of Tabenbuluk, western Kansu.' *Palaeontologia Sinica,* ser. C, no. **8a**, 1–113, **8b**, 1–259.

Brink, A. S. (1956) Speculations on some advanced mammalian characteristics in the higher mammal-like reptiles. *Palaeontologia Africana,* vol. **4**; 77–96.

Brinkmann, R. (1967) *Abriss der Geologie,* 2 vols. Stuttgart.

Bubnoff, S. von (1956) *Einführung in die Erdgeschichte.* Berlin.

Butzer, K. (1964) *Environment and Archaeology.* Chicago.

Chappell, W. M., Durham, J. W. & Savage, D. E. (1951) 'Mold of a rhinoceros in basalt, lower Grand Coulee, Washington.' *Bull. Geol. Soc. America,* **82**: 907–18.

Charlesworth, J. K. (1957) *The Quaternary era.* 2 vols. London.

Colbert, E. H. (1935) 'Siwalik mammals in the American Museum of Natural History.' *Trans. Amer. Philos. Soc.,* **26**: 1–401.
(1938) 'Fossil mammals from Burma in the American Museum of Natural History.' *Bull. Amer. Mus. Nat. Hist.,* **54**: 255–436.
(1965) *The age of reptiles.* London.
(1969) *Evolution of the vertebrates.* New York.

Cooke, H. B. S. (1967) 'The Pleistocene sequence in South Africa and problems of correlation.' In: Bishop & Clark (ed.): *Background to evolution in Africa,* 175–84. Chicago.

Crusafont, M., Truyols, J. & Villalta, J. F. (1955) 'El Burdigaliense continental de la cuenca del Vallés-Penedés.' *Mem. Comun. Inst. Geol. Dip. Prov.* Barcelona, **12**.

REFERENCES

Dehm, R. & Oettingen-Spielberg, T. (1958) 'Paläontologische und geologische Untersuchungen im Tertiär von Pakistan. 2. Die mitteleocänen Säugetiere von Ganda Kas bei Basal in Nordwest-Pakistan.' *Abh. Bayer. Akad. Wiss.*, **91**: 1–54.

Depéret, C. (1917) 'Monographie de la faune de mammifères fossiles du Ludien inférieur d'Euzet-les-Bains (Gard).' *Ann. Univ. Lyon*, n. ser. 1, **40**: 1–288.

Dietz, R. S. & Holden, J. C. (1970) 'The breakup of Pangaea.' *Scient. American*, 1970, October, 30–41.

Downs, T. (1956) 'The Mascall fauna from the Miocene of Oregon.' *Univ. California Publ. Geol. Sci.*, 31(5): 199–354.

Dunbar, C. O. (1949) *Historical geology*. New York.

Emiliani, C. (1958) 'Ancient temperatures.' *Sci. American*, 1958, February, 11 p.
(1961) 'Cenozoic climatic changes as indicated by the stratigraphy and chronology of deep-sea cores of Globigerina-ooze facies.' *Ann. New York Acad. Sci.*, 95: 521–36.
(1966) 'Paleotemperature analysis of Caribbean cores P6304–8 and P6304–9 and a generalised temperature curve for the past 425,000 years.' *Jour. Geol.*, **74**: 109–26.

Evernden, J. F. & James, G. T. (1964) 'Potassium-argon dates and the Tertiary floras of North America.' *Amer. Jour. Sci.*, **262**: 945–74.

Evernden, J. F., Savage, D. E., Curtis, G. H. & James, G. T. (1964) 'Potassium-argon dates and the Cenozoic mammalian chronology of North America.' *Amer. Jour. Sci.*, **262**: 145–98.

Filhol, H. (1876–77) 'Recherches sur les phosphorites du Quercy: étude des fossiles qu'on y rencontre et spécialment des mammifères.' *Ann. Sci. Géol.*, **7**: 1–220, **8**: 1–340.
(1882) 'Étude des mammifères fossiles de Ronzon (Haute-Loire).' *Ann. Sci. Géol.*, **12**: 1–271.

Flint, R. F. (1957) *Glacial and Pleistocene geology*. New York.

Gaudry, A. (1867) *Animaux fossiles et géologie de l'Attique*. 2 vols. Paris.

Gazin, C. L. (1956) 'Paleocene mammalian faunas of the Bison Basin in south-central Wyoming.' *Smithsonian Misc. Coll.*, 131 (6): 1–57.
(1962) 'A further study of the Lower Eocene mammalian faunas of southwestern Wyoming.' *Smithsonian Misc. Coll.*, 144 (1): 1–98.

Gromova, Vera (1954) (Swamp rhinoceroses (Amynodontidae) from Mongolia.) *Trudy Paleont. Inst. Akad. Nauk. SSSR.*, Moskva, **55**: 85–189.

Hildebrand, M. (1958) 'Cosmic events and perspective in organic evolution.' *Turtox News*, **36**: 112–13.

Hopkins, D. M. (1967) (ed.) *The Bering land bridge*. Stanford.

Howell, F. C. (1969) 'Remains of Hominidae from Pliocene/Pleistocene formations in the Lower Omo Basin, Ethiopia.' *Nature*, **223**: 1234–9.

Howells, W.W. (1967) *Mankind in the making. The story of human evolution.* Harmondsworth.

Huxley, J. (1942) *Evolution: the modern synthesis.* London.

Irving, E. (1964) *Paleomagnetism and its application to geological and geophysical problems.* New York.

Jepsen, G.L. (1940) 'Paleocene faunas of the Polecat Bench Formation, Park County, Wyoming. I.' *Proc. Amer. Philos. Soc.,* **83**: 217–340.
(1966) 'Early Eocene bat from Wyoming.' *Science,* **154**: 1333–9.

Jepsen, G.L. & Woodburne, M.O. (1969) 'Paleocene hyracothere from Polecat Bench Formation, Wyoming.' *Science,* **164**: 543–7.

Kellogg, R. (1936) 'A review of the Archaeoceti.' *Publ. Carnegie Inst. Washington,* **482**: 1–366.

Kermack, K.A. (1965) 'The origin of mammals.' *Science Jour.,* 1965, September, 70–2.

Krumbiegel, G. (1959) *Die tertiäre Pflanzen- und Tierwelt der Braunkohle des Geiseltales.* Wittenberg Lutherstadt.

Kröschke, O. (1936) 'Die Moa-Strausse, Neuseelands ausgestorbene Riesenvögel.' Wittenberg Lutherstadt.

Kuhn-Schnyder, E. (1953) *Geschichte der Wirbeltiere.* Basel.

Kummel, B. (1961) *History of the earth. An introduction to historical geology.* San Francisco.

Kurtén, B. (1965) 'The Carnivora of the Palestine caves.' *Acta Zool. Fennica,* **107**: 1–74.
(1966) 'Holarctic land connexions in the early Tertiary. Comment.' *Biol. Soc. Sci. Fennica,* **29** (5): 1–5.
(1968 a) *The age of the dinosaurs.* London.
(1968 b) *Pleistocene mammals of Europe.* London.
(1969) 'Continental drift and evolution.' *Scient. American,* 1969, March, 54–64.
(In the press) *The Ice Age.* London.

Lavocat, R. (1951) *Révision de la faune des mammifères oligocènes d'Auvergne et du Velay.* Paris.

Leakey, L.S.B. (1965) (ed.) *Olduvai Gorge 1951–1961.* Cambridge.

Le Gros Clark, W. (1964) *The fossil evidence for human evolution.* Chicago.
(1967) *Man-apes or ape-men? The story of discoveries in Africa.* New York.

Loomis, F.B. (1914) *The Deseado Formation of Patagonia.* Amherst.

Lyell, C. (1875) *The principles of geology.* 2 vols. London.

McKenna, M.C. (1960) 'Fossil Mammalia from the early Wasatchian Four Mile fauna, Eocene of northwest Colorado.' *Univ. California Publ. Geol. Sci.,* **37**: (L) 1–130.

REFERENCES

Mägdefrau, K. (1968) *Paläobiologie der Pflanzen.* Jena.

Martin, P. S. & Wright, H. E. (1967) (ed.) *Pleistocene extinctions. The search for a cause.* New Haven.

Matthew, W. D. (1924) 'Third contribution to the Snake Creek fauna.' *Bull. Amer. Mus. Nat. Hist.,* **50**: 59–210.
(1937) 'Paleocene faunas of the San Juan Basin, New Mexico.' *Trans. Amer. Philos. Soc.,* n.s., **30**: 1–510.

Matthew, W. D. & Granger, W. (1915–8) 'A revision of the Lower Eocene Wasatch and Wind River faunas. 1–5.' *Bull. Amer. Mus. Nat. Hist.,* **34**: 1–103, 311–61, 420–83; **38**: 565–657.
(1925 a) 'Fauna and correlation of the Gashato Formation of Mongolia.' *Amer. Mus. Novitates,* **186**: 1–12.
(1925 b) 'New mammals from the Indian Manha Eocene of Mongolia.' *Amer. Mus. Novitates,* **198**: 1–10.

Menard, H. W. (1969) 'The deep-ocean floor.' *Scient. American,* 1969, September, 126–42.

Merrilees, D. (1968) 'Man the destroyer: late Quaternary changes in the Australian marsupial fauna.' *Jour. Roy. Soc. W. Australia,* **51**: 1–24.

Mitchell, E. (1966) 'Faunal succession of extinct North Pacific marine mammals.' *Norsk Hvalfangst-Tidende,* 1966 (3): 47–60.

Nairn, A. R. M. (1961) (ed.) *Descriptive palaeoclimatology.* New York.

Oakley, K. P. (1964) *Frameworks for dating fossil man.* London.

Osborn, H. F. (1910) *The Age of Mammals in Europe, Asia and North America.* New York.

Papp, A. & Thenius, E. (1959) *Tertiär I.* Stuttgart.

Paula Couto, C. de (1952) 'Fossil mammals from the beginning of the Cenozoic in Brazil. Condylarthra, Litopterna, Xenungulata and Astrapotheria.' *Bull. Amer. Mus. Nat. His.,* **99** (6): 355–94.

Pearson, R. (1964) *Animals and plants of the Cenozoic era.* London.

Pei, W. C. (1934) 'On the Carnivora from locality 1 of Choujoutien.' *Palaeontologia Sinica, ser.* C, **8** (1): 1–216.

Pilbeam, D. (1966) 'Notes on *Ramapithecus,* the earliest known hominid, and *Dryopithecus.' Amer. Jour. Phys. Anthrop.,* n.s., **25**: 1–6.

Ride, W. D. L. (1964) 'A review of Australian fossil marsupials.' *Jour. Roy. Soc. W. Australia,* **47**: 97–131.

Romer, A. S. (1967) *Vertebrate paleontology.* Chicago.
(1968) *Notes and comments on vertebrate paleonthology.* Chicago.

Runcorn, S.K.(1962) (ed.) *Continental drift*. London.

Russell, D.E.(1962) 'Essai de reconstitution de la vie paléocène au Mont de Berru.' *Null. Mus. Natl. Hist. Nat. Paris*, ser. 2, **34**: 101–6.
(1964) 'Les mammifères Paléocènes d'Europe.' *Mém. Mus. Natl. Hist. Nat. Paris*, n.s., **13**: 1–321.

Savage, D.E.(1951) 'Late Cenozoic vertebrates of the San Francisco Bay region.' *Univ. California Publ. Geol. Sci.*, **28** (10): 215–314.

Schlosser, M.(1911) 'Beiträge zur Kenntnis der oligozänen Landsäugetiere aus dem Fayum, Agypten.' *Beitr. Pal. Geol. Osterreich-Ungarn*, **24**: 51–167.

Schuchert, C.(1955) *Atlas of paleogeographic maps of North America*. New York.

Scott, W.B.(1912) 'Mammalia of the Santa Cruz beds. Entelonychia; Order Toxodontia.' *Princeton Univ. Exp. Patagonia*, **6**: 239–300.
(1962) *A history of land mammals in the Western hemisphere*. New York. (Reprint; original 1937.)

Scott, W. B., Jepsen, G. L. & Wood, A. E.(1941) 'The mammalian fauna of the White River Oligocene.' *Trans. Amer. Philos. Soc.* N.S. 28: 1–980.

Simons, E.L.(1967) 'The earliest apes.' *Scient. American*, 1967, December, 28–35.

Simpson, G.G.(1937) 'The Fort Union of the Crazy Mountain field, Montana, and its mammalian fauna.' *Bull. U.S. Natl. Mus.*, **169**: 1–287.
(1945) 'The principles of classification and a classification of mammals.' *Bull. Amer. Mus. Nat. Hist.*, **85**: 1–450.
(1953) *Life of the past*. New Haven.
(1960) 'Man's evolutionary future.' *Zool. Jahrbücher*, **88**: 125–34.
(1961) 'Historical zoogeography of Australian mammals.' *Evolution*, **15**: 431–46.
(1962) 'Evolution of Mesozoic mammals.' *Internatl. Colloquium on Evolution of Mammals*, **1**: 57–95. Brussels.
(1967) 'The beginning of the age of mammals in South America. II.' *Bull. Amer. Mus. Nat. Hist.*, **137**: 1–259.

Sloan, R.E.& Van Valen, L.(1965) 'Cretaceous mammals from Montana.' *Science*, **148**: 220–7.

Stehlin, H.G.(1903–16) 'Die säugetiere des schweizerischen Eocaens: critischer Catalog der Materialien.' *Abh. Schweiz. Palaeont. Ges.*, **30**: 1–153; **31**: 155–455; **32**: 447–595; **33**: 597;690; **35**: 691–837; **36**: 839–1164; **38**: 1165–1298; **41**: 1297–1552.

Stock, C.(1930) 'Rancho La Brea: a record of Pleistocene life in California (also later editions).' *Publ. Los Angeles County Mus.*, **1**: 1–81.

Swinton, W.E.(1962) *Fossil amphibians and reptiles*. London.

Teilhard de Chardin, P.(1922) 'Les mammifères de l'Eocène inférieur français et leurs gisements.' *Ann. Paléont.*, **11**: 9–116.

REFERENCES

Thenius, E. (1952) 'Die Säugetierfauna aus dem Torton von Neudorf an der March (CSR).' *Neues Jahrb. Min. Geol. Palaont.*, **96**: 27–136.
(1959) *Tertiär II. Wirbeltierfaunen.* Stuttgart.

Thenius, E. & Hofer, H. (1960) *Stammesgeschichte der Säugetiere. Eine Ubersicht über Tatsachen und Probleme der Evolution der Säugetiere.* Berlin.

Toepfer, V. (1963) *Tierwelt des Eiszeitalters.* Leipzig.

Toots, H. & Voorhies, M. R. (1965) 'Strontium in fossil bones and the reconstruction of food chains.' *Science*, **149**: 854–5.

Wills, L. J. (1952) *A palaeogeographical atlas of the British Isles and adjacent parts of Europe.* London.

Wilson, R. W. (1960) 'Early Miocene rodents and insectivores from northeastern Colorado.' *Univ. Kansas Publ., Palaeont. Contr., Vertebrata*, **7**: 1–92.

Woldstedt, P. (1969) *Quartär.* Stuttgart.

Woodburne, M. O. (1967) 'The Alcoota fauna, Central Australia. An integrated palaeontological and geological study.' *Bull. Bur. Min. Res. Australia*, **87**: 1–187.

Woodford, A. O. (1965) *Historical geology.* San Francisco.

Wright, H. E. & Frey, D. G. (1965) (ed.) *The Quaternary of the United States.* Princeton.

Zagwijn, W. H. (1967) 'Ecologic interpretation of a pollen diagram from Neocene beds in the Netherlands.' *Rev. Palaeobotany & Palynology*, **2**: 173–81.

Zeuner, F. E. (1959) *The Pleistocene period.* London.
(1961) *Dating the past.* London.

Zittel, K. A. (1923) *Grundzüge der Paläontologie.* Munich.

Index to Authors

Index to stratigraphic and locality names

Index to Latin names